Learning math can be an exciting adventure.
That's why *Illustrative Mathematics* is designed
to provide opportunities for you to learn math
by doing math. By engaging in problem-solving
activities, you will make and test your own
conclusions and try out different strategies
as you learn math.

Let's get started!

Unit 1
Rigid Transformations and Congruence

Contents in Brief

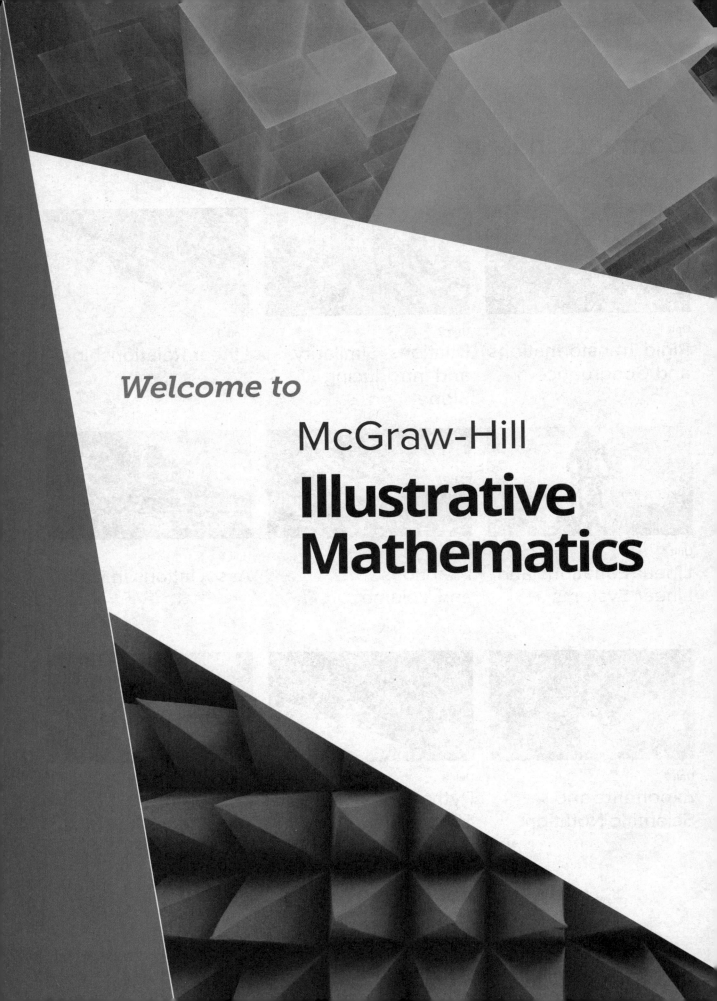

Welcome to

McGraw-Hill
Illustrative Mathematics

Unit 2

Dilations, Similarity, and Introducing Slope

Unit 3

Linear Relationships

Unit 4

Linear Equations and Linear Systems

Unit 5

Functions and Volume

Unit 6

Associations in Data

Unit 7

Exponents and Scientific Notation

Unit 8

Pythagorean Theorem and Irrational Numbers

Unit 9

Putting It All Together

Unit 1

Rigid Transformations and Congruence

The lake shows a mirror image, or a reflection, of the mountain scene. You'll learn more about reflections in this unit.

Topics
- Rigid Transformations
- Properties of Rigid Transformations
- Congruence
- Angles in a Triangle
- Let's Put It to Work

Rigid Transformations and Congruence

Lesson 1-1

Moving in the Plane

NAME _____ DATE _____ PERIOD _____

Learning Goal Let's describe ways figures can move in the plane.

 ## Warm Up
1.1 Which One Doesn't Belong: Diagrams

Which one doesn't belong?

Diagram A   **Diagram B**   **Diagram C**   **Diagram D**

 ## Activity
1.2 Triangle Square Dance

Your teacher will give you three pictures. Each shows a different set of dance moves.

1. Arrange the three pictures so you and your partner can both see them right way up. Choose who will start the game.

   • The starting player mentally chooses A, B, or C and describes the dance to the other player.

   • The other player identifies which dance is being talked about: A, B, or C.

2. After one round, trade roles. When you have described all three dances, come to an agreement on the words you use to describe the moves in each dance.

3. With your partner, write a description of the moves in each dance.

Are you ready for more?

We could think of each dance as a new dance by running it in reverse, starting in the 6th frame and working backwards to the first.

1. Pick a dance and describe in words one of these reversed dances.

2. How do the directions for running your dance in the forward direction and the reverse direction compare?

Summary

Moving in the Plane

Here are two ways for changing the position of a figure in a plane without changing its shape or size.

1. Sliding or shifting the figure without turning it. Shifting Figure A to the right and up puts it in the position of Figure B.

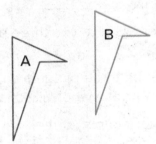

2. Turning or rotating the figure around a point. Figure A is rotated around the bottom **vertex** to create Figure C.

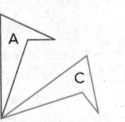

Glossary
_____

vertex

NAME _____ DATE _____ PERIOD _____

Practice
Moving in the Plane

1. The six frames show a shape's different positions.

   Describe how the shape moves to get from its position in each frame to the next.

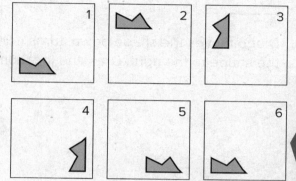

2. These five frames show a shape's different positions.

   Describe how the shape moves to get from its position in each frame to the next.

3. Diego started with this shape.

Diego moves the shape down, turns it 90 degrees clockwise, then moves the shape to the right. Draw the location of the shape after each move.

Lesson 1-2

Naming the Moves

NAME _____ DATE _____ PERIOD _____

Learning Goal Let's be more precise about describing moves of figures in the plane.

Warm Up
2.1 A Pair of Quadrilaterals

Quadrilateral A can be rotated into the position of Quadrilateral B. Estimate the angle of rotation.

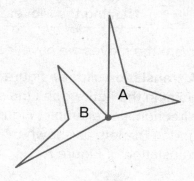

Activity
2.2 How Did You Make That Move?

Here is another set of dance moves.

1. Describe each move or say if it is a new move.

   a. Frame 1 to Frame 2.

   b. Frame 2 to Frame 3.

   c. Frame 3 to Frame 4.

   d. Frame 4 to Frame 5.

   e. Frame 5 to Frame 6.

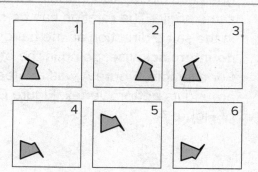

2. How would you describe the new move?

Your teacher will give you a set of cards. Sort the cards into categories according to the type of move they show. Be prepared to describe each category and why it is different from the others.

Summary
Naming the Moves

Here are the moves we have learned about so far:

- A **translation** slides a figure without turning it. Every point in the figure goes the same distance in the same direction. For example, Figure A was translated down and to the left, as shown by the arrows. Figure B is a translation of Figure A.

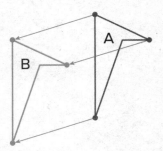

- A **rotation** turns a figure about a point, called the center of the rotation. Every point on the figure goes in a circle around the center and makes the same angle. The rotation can be **clockwise**, going in the same direction as the hands of a clock, or **counterclockwise**, going in the other direction. For example, Figure A was rotated 45° clockwise around its bottom vertex. Figure C is a rotation of Figure A.

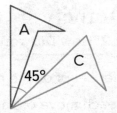

NAME _____ DATE _____ PERIOD _____

- A **reflection** places points on the opposite side of a reflection line. The mirror image is a backwards copy of the original figure. The reflection line shows where the mirror should stand. For example, Figure A was reflected across the dotted line. Figure D is a reflection of Figure A.

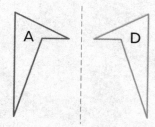

We use the word *image* to describe the new figure created by moving the original figure. If one point on the original figure moves to another point on the new figure, we call them *corresponding* points.

Glossary

clockwise

counterclockwise

reflection

rotation

translation

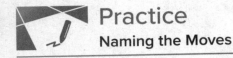

Practice
Naming the Moves

1. Each of the six cards shows a shape.

   a. Which pair of cards shows a shape and its image after a rotation?

   b. Which pair of cards shows a shape and its image after a reflection?

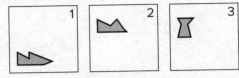

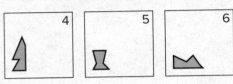

2. The five frames show a shape's different positions. Describe how the shape moves to get from its position in each frame to the next.

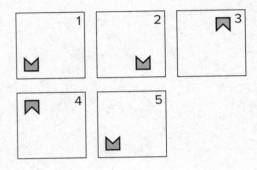

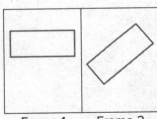

3. The rectangle seen in Frame 1 is rotated to a new position, seen in Frame 2.

   Select **all** the ways the rectangle could have been rotated to get from Frame 1 to Frame 2. **(Lesson 1-1)**

   A. 40 degrees clockwise

   B. 40 degrees counterclockwise

   C. 90 degrees clockwise

   D. 90 degrees counterclockwise

   E. 140 degrees clockwise

   F. 140 degrees counterclockwise

Lesson 1-3

Grid Moves

NAME _____ DATE _____ PERIOD _____

Learning Goal Let's transform some figures on grids.

Warm Up
3.1 Notice and Wonder: The Isometric Grid

What do you notice?
What do you wonder?

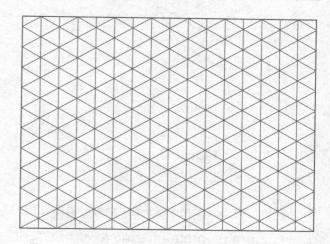

Activity
3.2 Transformation Information

Your teacher will give you tracing paper to carry out the moves specified. Use A′, B′, C′, and D′ to indicate vertices in the new figure that correspond to the points A, B, C, and D in the original figure.

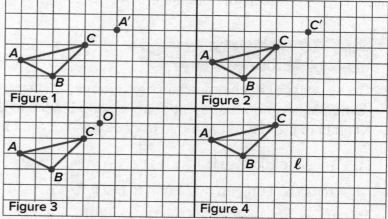

1. In Figure 1, translate triangle *ABC* so that *A* goes to *A′*.

2. In Figure 2, translate triangle *ABC* so that *C* goes to *C′*.

3. In Figure 3, rotate triangle *ABC* 90° counterclockwise using center *O*.

4. In Figure 4, reflect triangle *ABC* using line ℓ.

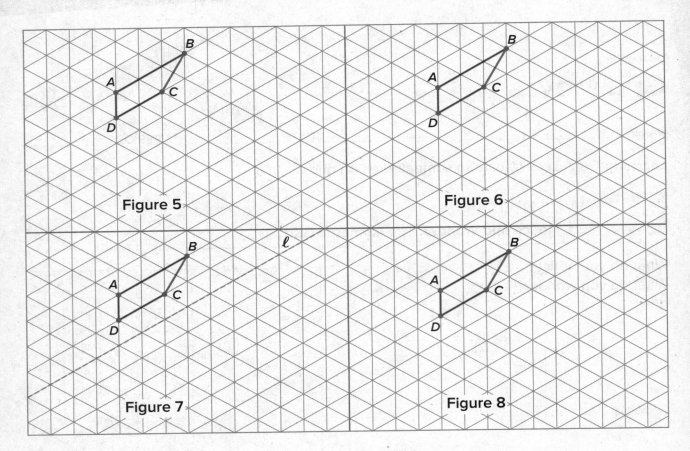

Figure 5

Figure 6

Figure 7

Figure 8

5. In Figure 5, rotate quadrilateral *ABCD* 60° counterclockwise using center *B*.

6. In Figure 6, rotate quadrilateral *ABCD* 60° clockwise using center C.

7. In Figure 7, reflect quadrilateral *ABCD* using line ℓ.

8. In Figure 8, translate quadrilateral *ABCD* so that *A* goes to *C*.

Are you ready for more?

The effects of each move can be "undone" by using another move. For example, to undo the effect of translating 3 units to the right, we could translate 3 units to the left. What move undoes each of the following moves?

1. Translate 3 units up

2. Translate 1 unit up and 1 unit to the left

3. Rotate 30 degrees clockwise around a point *P*

4. Reflect across a line ℓ

NAME _____ DATE _____ PERIOD _____

Summary
Grid Moves

When a figure is on a grid, we can use the grid to describe a transformation. For example, here is a figure and an **image** of the figure after a move.

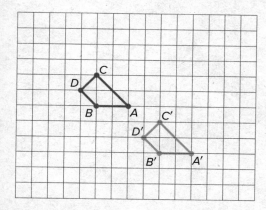

Quadrilateral *ABCD* is translated 4 units to the right and 3 units down to the position of quadrilateral *A'B'C'D'*.

A second type of grid is called an *isometric grid*. The isometric grid is made up of equilateral triangles. The angles in the triangles all measure 60 degrees, making the isometric grid convenient for showing rotations of 60 degrees.

Here is quadrilateral *KLMN* and its image *K'L'M'N'* after a 60-degree counterclockwise rotation around a point *P*.

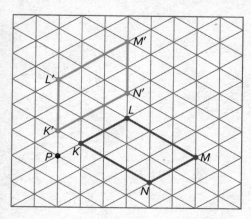

Glossary

image

1. Apply each transformation described to Figure A. If you get stuck, try using tracing paper.

   a. A translation which takes P to P'

   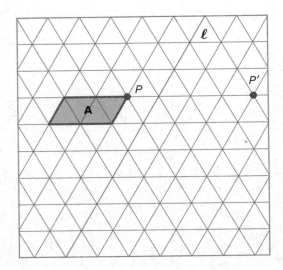

   b. A counterclockwise rotation of A, using center P, of 60 degrees

   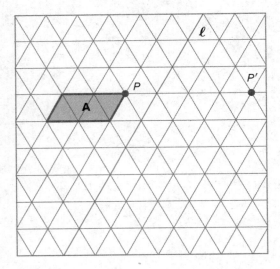

NAME _____ DATE _____ PERIOD _____

c. A reflection of A across line ℓ

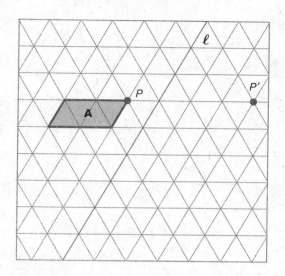

2. Here is triangle *ABC* drawn on a grid.

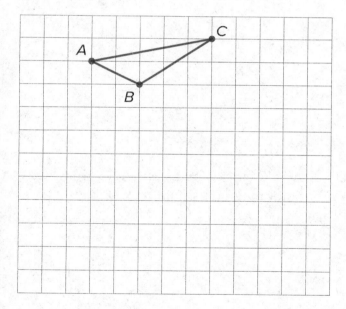

On the grid, draw a rotation of triangle *ABC*, a translation of triangle *ABC*, and a reflection of triangle *ABC*. Describe clearly how each was done.

3. Respond to each of the following. (Lesson 1-2)

a. Draw the translated image of *ABCDE* so that vertex *C* moves to *C'*. Tracing paper may be useful.

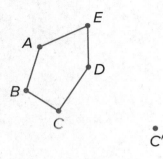

b. Draw the reflected image of Pentagon *ABCDE* with line of reflection ℓ. Tracing paper may be useful.

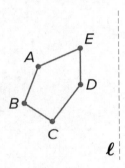

c. Draw the rotation of Pentagon *ABCDE* around *C* clockwise by an angle of 150 degrees. Tracing paper and a protractor may be useful.

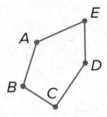

Lesson 1-4

Making the Moves

NAME _____ DATE _____ PERIOD _____

Learning Goal Let's draw and describe translations, rotations, and reflections.

Warm Up
4.1 Reflection Quick Image

Here is an incomplete image. Your teacher will display the completed image twice, for a few seconds each time. Your job is to complete the image on your copy.

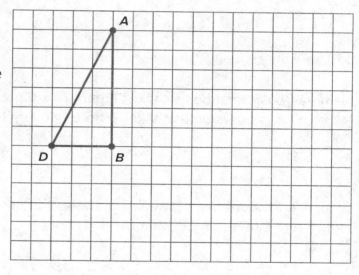

Activity
4.2 Make That Move

Your partner will describe the image of this triangle after a certain **transformation**. Sketch it here.

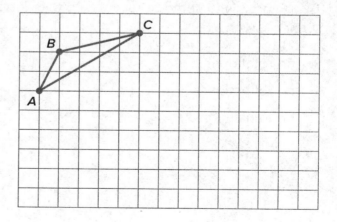

Activity

4.3 A to B to C

Here are some figures on an isometric grid.

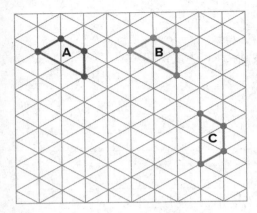

1. Name a transformation that takes Figure A to Figure B.
   Name a transformation that takes Figure B to Figure C.

2. What is one **sequence of transformations** that takes
   Figure *A* to Figure *C*? Explain how you know.

Are you ready for more?

Experiment with some other ways to take Figure *A* to Figure *C*. For example, can
you do it with. . .

- no rotations?

- no reflections?

- no translations?

NAME _____ DATE _____ PERIOD _____

Summary
Making the Moves

A move, or combination of moves, is called a **transformation**. When we do one or more moves in a row, we often call that a **sequence of transformations**. To distinguish the original figure from its image, points in the image are sometimes labeled with the same letters as the original figure, but with the symbol ′ attached, as in *A′* (pronounced "A prime").

- A translation can be described by two points. If a translation moves point *A* to point *A′*, it moves the entire figure the same distance and direction as the distance and direction from *A* to *A′*. The distance and direction of a translation can be shown by an arrow.

  For example, here is a translation of quadrilateral *ABCD* that moves *A* to *A′*.

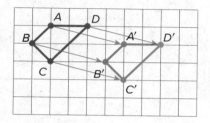

- A rotation can be described by an angle and a center. The direction of the angle can be clockwise or counterclockwise.

  For example, hexagon *ABCDEF* is rotated 90° counterclockwise using center *P*.

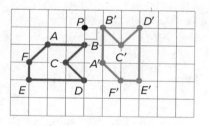

- A reflection can be described by a line of reflection (the "mirror"). Each point is reflected directly across the line so that it is just as far from the mirror line, but is on the opposite side.

  For example, pentagon *ABCDE* is reflected across line *m*.

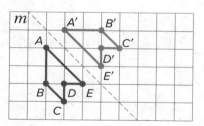

Glossary

sequence of transformations
transformation

Practice
Making the Moves

1. For each pair of polygons, describe a sequence of translations, rotations, and reflections that takes Polygon P to Polygon Q.

a.

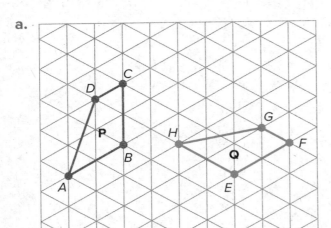

b.

NAME _____ DATE _____ PERIOD _____

c.

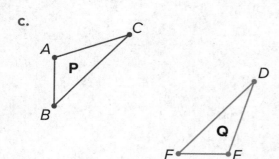

2. Here is quadrilateral *ABCD* and line ℓ.

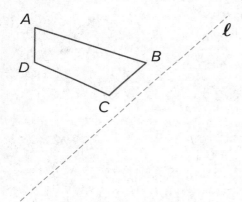

Draw the image of quadrilateral *ABCD* after reflecting it across line ℓ. **(Lesson 1-2)**

3. Here is quadrilateral *ABCD*.
Draw the image of quadrilateral *ABCD* after each rotation using
B as center. **(Lesson 1-2)**

   a. 90 degrees clockwise

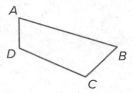

   b. 120 degrees clockwise

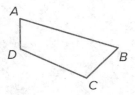

   c. 30 degrees counterclockwise

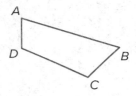

Lesson 1-5

Coordinate Moves

NAME _____ DATE _____ PERIOD _____

Learning Goal Let's transform some figures and see what happens to the coordinates of points.

 ## Warm Up
5.1 Translating Coordinates

Select **all** of the translations that take Triangle T to Triangle U.
There may be more than one correct answer.

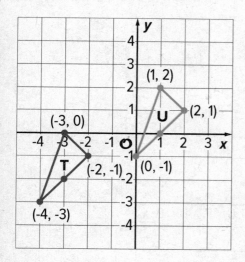

(A.) Translate (-3, 0) to (1, 2).

(B.) Translate (2, 1) to (-2, -1).

(C.) Translate (-4, -3) to (0, -1).

(D.) Translate (1, 2) to (2, 1).

1. Here is a list of points

   $A = (0.5, 4)$ $B = (-4, 5)$ $C = (7, -2)$ $D = (6, 0)$ $E = (0, -3)$

   On the **coordinate plane**:

   a. Plot each point and label each with its coordinates.

   b. Using the *x*-axis as the line of reflection, plot the image of each point.

   c. Label the image of each point with its coordinates.

   d. Include a label using a letter. For example, the image of point *A* should be labeled *A'*.

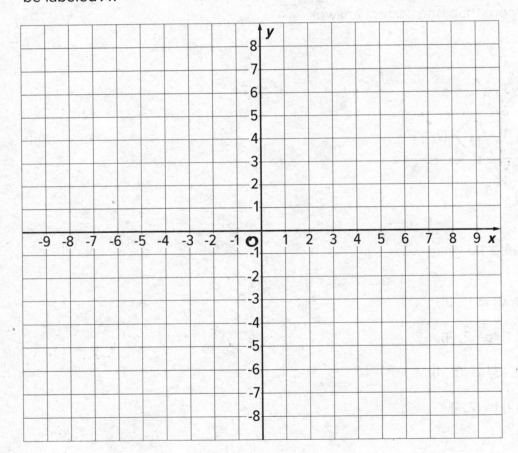

NAME _____ DATE _____ PERIOD _____

2. If the point (13, 10) were reflected using the *x*-axis as the line of reflection, what would be the coordinates of the image? What about (13, -20)? (13, 570)? Explain how you know.

3. The point *R* has coordinates (3, 2).

   a. Without graphing, predict the coordinates of the image of point *R* if point *R* were reflected using the *y*-axis as the line of reflection.

   b. Check your answer by finding the image of *R* on the graph.

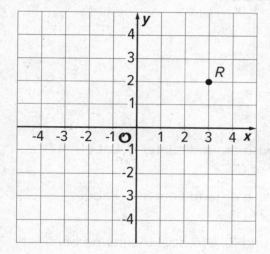

   c. Label the image of point *R* as *R'*.

   d. What are the coordinates of *R'*?

4. Suppose you reflect a point using the *y*-axis as the line of reflection. How would you describe its image?

Activity

5.3 Transformations of a Segment

Apply each of the following transformations to segment *AB*.

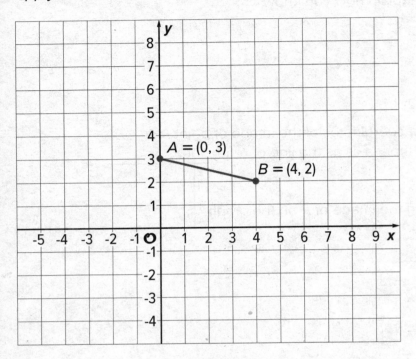

1. Rotate segment *AB* 90 degrees counterclockwise around center *B*. Label the image of *A* as *C*. What are the coordinates of *C*?

2. Rotate segment *AB* 90 degrees counterclockwise around center *A*. Label the image of *B* as *D*. What are the coordinates of *D*?

3. Rotate segment *AB* 90 degrees clockwise around (0, 0). Label the image of *A* as *E* and the image of *B* as *F*. What are the coordinates of *E* and *F*?

4. Compare the two 90-degree counterclockwise rotations of segment *AB*. What is the same about the images of these rotations? What is different?

Are you ready for more?

Suppose *EF* and *GH* are line segments of the same length. Describe a sequence of transformations that moves *EF* to *GH*.

NAME _____ DATE _____ PERIOD _____

Summary
Coordinate Moves

We can use coordinates to describe points and find patterns in the coordinates of transformed points.

We can describe a translation by expressing it as a sequence of horizontal and vertical translations. For example, segment *AB* is translated right 3 and down 2.

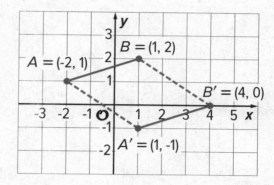

Reflecting a point across an axis changes the sign of one coordinate. For example,

• Reflecting the point *A* whose coordinates are (2, -1) across the *x*-axis changes the sign of the *y*-coordinate, making its image the point *A'* whose coordinates are (2, 1).

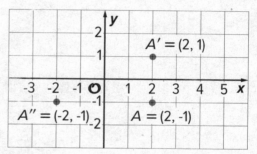

• Reflecting the point *A* across the *y*-axis changes the sign of the *x*-coordinate, making the image the point *A''* whose coordinates are (-2, -1).

Reflections across other lines are more complex to describe.

We don't have the tools yet to describe rotations in terms of coordinates in general. Here is an example of a 90° rotation with center (0, 0) in a counterclockwise direction.

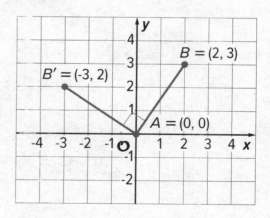

Point *A* has coordinates (0, 0). Segment *AB* was rotated 90° counterclockwise around *A*. Point *B* with coordinates (2, 3) rotates to point *B'* whose coordinates are (-3, 2).

Glossary

coordinate plane

Practice
Coordinate Moves

1. **a.** Here are some points.

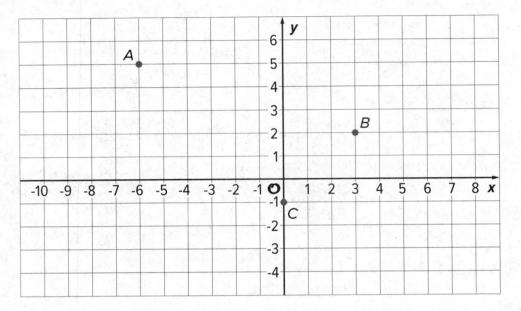

What are the coordinates of *A*, *B*, and *C* after a translation to the right by 4 units and up 1 unit? Plot these points on the grid, and label them *A′*, *B′* and *C′*.

b. Here are some points.

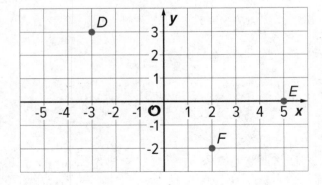

What are the coordinates of *D*, *E*, and *F* after a reflection over the *y* axis? Plot these points on the grid, and label them *D′*, *E′* and *F′*.

NAME _____ DATE _____ PERIOD _____

c. Here are some points.

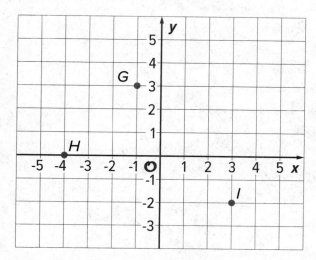

What are the coordinates of *G*, *H*, and *I* after a rotation about (0, 0) by 90 degrees clockwise? Plot these points on the grid, and label them *G′*, *H′* and *I′*.

2. Describe a sequence of transformations that takes trapezoid A to trapezoid B. **(Lesson 1-4)**

3. Reflect polygon P using line ℓ. (Lesson 1-3)

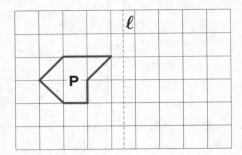

Lesson 1-6

Describing Transformations

NAME _____ DATE _____ PERIOD _____

Learning Goal Let's transform some polygons in the coordinate plane.

Warm Up
6.1 Finding a Center of Rotation

Andre performs a 90-degree counterclockwise rotation of Polygon P and gets Polygon P′, but he does not say what the center of the rotation is. Can you find the center?

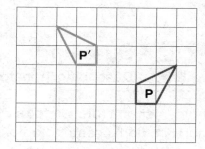

Activity
6.2 Info Gap: Transformation Information

Your teacher will give you either a *problem card* or a *data card*. Do not show or read your card to your partner.

If you have a *problem card*...	If you have a *data card*...

1. Silently read your card and think about what information you need to be able to answer the question.
 	
2. Ask your partner for the specific information that you need.
 	
3. Explain how you are using the information to solve the problem.
 	
  Continue to ask questions until you have enough information to solve the problem.
 	
4. Share the *problem card* and solve the problem independently.
 	
5. Read the *data card* and discuss your reasoning.
	1. Silently read your card.
	
	2. Ask your partner *"What specific information do you need?"* and wait for them to *ask* for information.
	
	  If your partner asks for information that is not on the card, do not do the calculations for them. Tell them you don't have that information.
	
	3. Before sharing the information, ask "Why do you need that information?" Listen to your partner's reasoning and ask clarifying questions.
	
	4. Read the *problem card* and solve the problem independently.
	
	5. Share the *data card* and discuss your reasoning.

Pause here so your teacher can review your work. Ask your teacher for a new set of cards and repeat the activity, trading roles with your partner.

Sometimes two transformations, one performed after the other, have a nice description as a single transformation. For example, instead of translating 2 units up followed by translating 3 units up, we could simply translate 5 units up. Instead of rotating 20 degrees counterclockwise around the origin followed by rotating 80 degrees clockwise around the origin, we could simply rotate 60 degrees clockwise around the origin.

Can you find a simple description of reflecting across the x-axis followed by reflecting across the y-axis?

Summary

Describing Transformations

The center of a rotation for a figure doesn't have to be one of the points on the figure. To find a center of rotation, look for a point that is the same distance from two corresponding points. You will probably have to do this for a couple of different pairs of corresponding points to nail it down.

When we perform a sequence of transformations, the order of the transformations can be important. Here is triangle *ABC* translated up two units and then reflected over the *x*-axis.

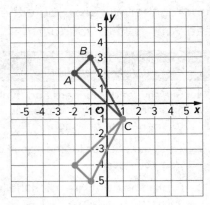

Here is triangle *ABC* reflected over the *x*-axis and then translated up two units.

Triangle *ABC* ends up in different places when the transformations are applied in the opposite order!

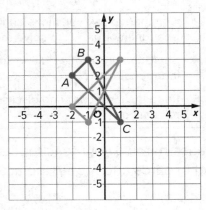

NAME _____ DATE _____ PERIOD _____

Practice
Describing Transformations

1. Here is Trapezoid A in the coordinate plane.

   a. Draw Polygon B, the image of A, using the *y*-axis as the line of reflection.

   b. Draw Polygon C, the image of B, using the *x*-axis as the line of reflection.

   c. Draw Polygon D, the image of C, using the *x*-axis as the line of reflection.

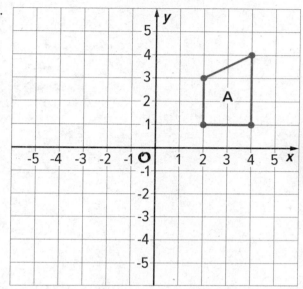

2. The point (-4, 1) is rotated 180 degrees counterclockwise using center (-3, 0). What are the coordinates of the image?

   A. (-5, -2)   **B.** (-4, -1)   **C.** (-2, -1)   **D.** (4, -1)

3. Describe a sequence of transformations for which Triangle B is the image of Triangle A.

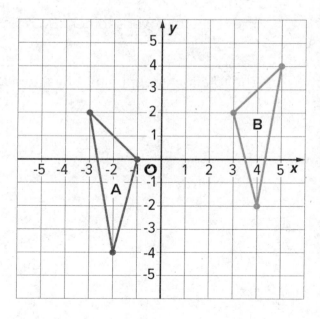

4. Here is quadrilateral *ABCD*. (Lesson 1-2)

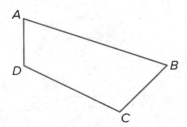

Draw the image of quadrilateral *ABCD* after each transformation.

a. The translation that takes *B* to *D*.

b. The reflection over segment *BC*.

c. The rotation about point *A* by angle *DAB*, counterclockwise.

Lesson 1-7

No Bending or Stretching

NAME _____ DATE _____ PERIOD _____

Learning Goal Let's compare measurements before and after translations, rotations, and reflections.

Warm Up
7.1 Measuring Segments

For each question, the unit is represented by the large tick marks with whole numbers.

1. Find the length of this segment to the nearest $\frac{1}{8}$ of a unit.

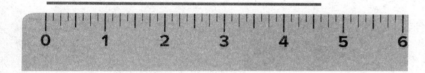

2. Find the length of this segment to the nearest 0.1 of a unit.

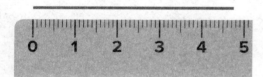

3. Estimate the length of this segment to the nearest $\frac{1}{8}$ of a unit.

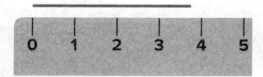

4. Estimate the length of the segment in the prior question to the nearest 0.1 of a unit.

Activity

7.2 Sides and Angles

1. Translate Polygon A so point P goes to point P'. In the image, write the length of each side, in grid units, next to the side.

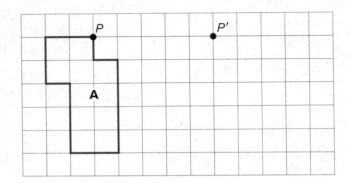

2. Rotate Triangle B 90 degrees clockwise using R as the center of rotation. In the image, write the measure of each angle in its interior.

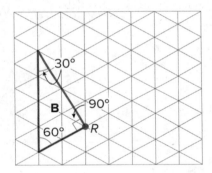

3. Reflect Pentagon C across line ℓ.

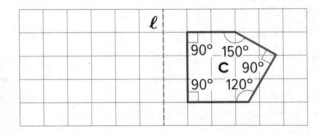

   a. In the image, write the length of each side, in grid units, next to the side. You may need to make your own ruler with tracing paper or a blank index card.

   b. In the image, write the measure of each angle in the interior.

NAME _____ DATE _____ PERIOD _____

Activity
7.3 Which One?

Here is a grid showing triangle *ABC* and two other triangles.

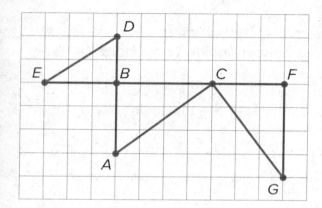

You can use a **rigid transformation** to take triangle *ABC* to *one* of the other triangles.

1. Which one? Explain how you know.

2. Describe a rigid transformation that takes *ABC* to the triangle you selected.

Are you ready for more?

A square is made up of an L-shaped region and three transformations of the region. If the perimeter of the square is 40 units, what is the perimeter of each L-shaped region?

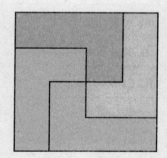

The transformations we've learned about so far, translations, rotations, reflections, and sequences of these motions, are all examples of **rigid transformations**. A rigid transformation is a move that doesn't change measurements on any figure.

Earlier, we learned that a figure and its image have corresponding points. With a rigid transformation, figures like polygons also have **corresponding** sides and corresponding angles. These corresponding parts have the same measurements. For example, triangle *EFD* was made by reflecting triangle *ABC* across a horizontal line, then translating. Corresponding sides have the same lengths, and corresponding angles have the same measures.

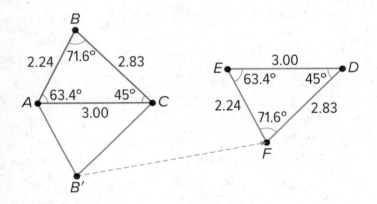

Measurements in Triangle *ABC*	Corresponding Measurements in Image *EFD*
AB = 2.24	*EF* = 2.24
BC = 2.83	*FD* = 2.83
CA = 3.00	*DE* = 3.00
m∠*ABC* = 71.6°	m∠*EFD* = 71.6°
m∠*BCA* = 45.0°	m∠*FDE* = 45.0°
m∠*CAB* = 63.4°	m∠*DEF* = 63.4°

Glossary

corresponding
rigid transformation

NAME _____ DATE _____ PERIOD _____

Practice
No Bending or Stretching

---

1. Is there a rigid transformation taking Rhombus P to Rhombus Q? Explain how you know.

2. Describe a rigid transformation that takes Triangle A to Triangle B.

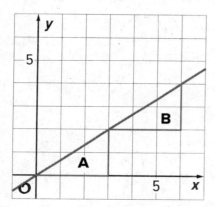

3. Is there a rigid transformation taking Rectangle A to Rectangle B? Explain how you know.

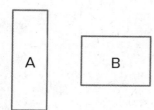

4. For each shape, draw its image after performing the transformation. If you get stuck, consider using tracing paper. **(Lesson 1-4)**

   a. Translate the shape so that A goes to A'.

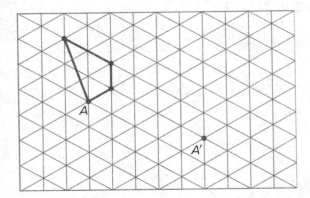

   b. Rotate the shape 180 degrees counterclockwise around B.

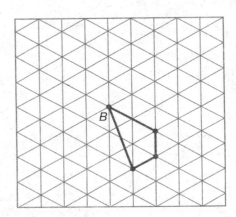

   c. Reflect the shape over the line shown.

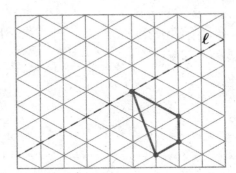

Lesson 1-8

Rotation Patterns

NAME _____ DATE _____ PERIOD _____

Learning Goal Let's rotate figures in a plane.

Warm Up
8.1 Building a Quadrilateral

Here is a right isosceles triangle.

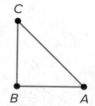

1. Rotate triangle *ABC* 90 degrees clockwise around *B*.

2. Rotate triangle *ABC* 180 degrees clockwise around *B*.

3. Rotate triangle *ABC* 270 degrees clockwise around *B*.

4. What would it look like when you rotate the four triangles 90 degrees clockwise around *B*? 180 degrees? 270 degrees clockwise?

Activity

8.2 Rotating a Segment

Refer to segment *CD* and point *E*.

1. Rotate segment *CD* 180 degrees around point *D*. Draw its image and label the image of *C* as *A*.

2. Rotate segment *CD* 180 degrees around point *E*. Draw its image and label the image of *C* as *B* and the image of *D* as *F*.

3. Rotate segment *CD* 180 degrees around its midpoint, *G*. What is the image of *C*?

4. What happens when you rotate a segment 180 degrees around a point?

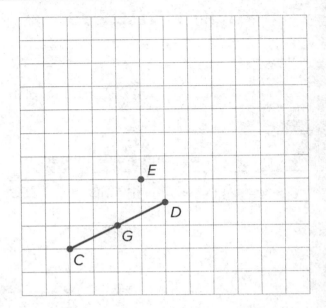

Are you ready for more?

Here are two line segments. Is it possible to rotate one line segment to the other? If so, find the center of such a rotation. If not, explain why not.

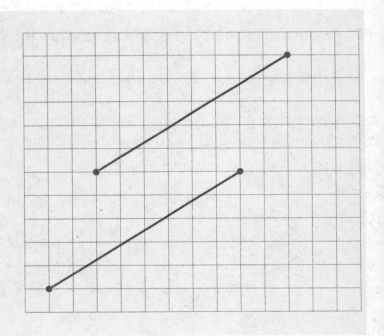

NAME _____ DATE _____ PERIOD _____

Activity

8.3 A Pattern of Four Triangles

You can use rigid transformations of a figure to make patterns. Here is a diagram built with three different transformations of triangle *ABC*.

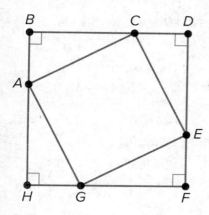

1. Describe a rigid transformation that takes triangle *ABC* to triangle *CDE*.

2. Describe a rigid transformation that takes triangle *ABC* to triangle *EFG*.

3. Describe a rigid transformation that takes triangle *ABC* to triangle *GHA*.

4. Do segments *AC*, *CE*, *EG*, and *GA* all have the same length? Explain your reasoning.

When we apply a 180-degree rotation to a line segment, there are several possible outcomes.

- The segment maps to itself (if the center of rotation is the midpoint of the segment).

- The image of the segment overlaps with the segment and lies on the same line (if the center of rotation is a point on the segment).

- The image of the segment does not overlap with the segment (if the center of rotation is *not* on the segment).

We can also build patterns by rotating a shape.

For example, triangle *ABC* shown here has $m(\angle A) = 60$. If we rotate triangle *ABC* 60 degrees, 120 degrees, 180 degrees, 240 degrees, and 300 degrees clockwise, we can build a hexagon.

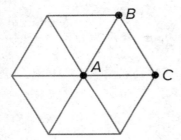

NAME _____ DATE _____ PERIOD _____

 Practice
Rotation Patterns

1. For the figure shown here,

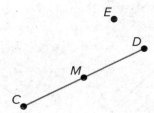

a. Rotate segment *CD* 180° around point *D*.

b. Rotate segment *CD* 180° around point *E*.

c. Rotate segment *CD* 180° around point *M*.

2. Here is an isosceles right triangle.

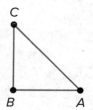

Draw these three rotations of triangle *ABC* together.

a. Rotate triangle *ABC* 90 degrees clockwise around *A*.

b. Rotate triangle *ABC* 180 degrees around *A*.

c. Rotate triangle *ABC* 270 degrees clockwise around *A*.

3. Each graph shows two polygons *ABCD* and *A'B'C'D'*. In each case, describe a sequence of transformations that takes *ABCD* to *A'B'C'D'*. (Lesson 1-5)

a.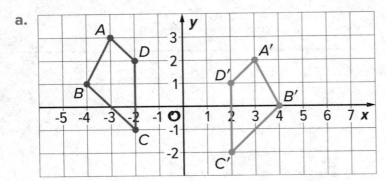

b.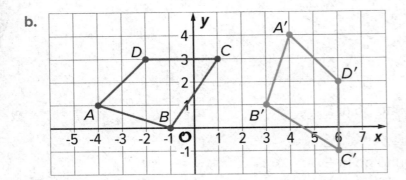

4. Lin says that she can map Polygon A to Polygon B using *only* reflections. Do you agree with Lin? Explain your reasoning. (Lesson 1-4)

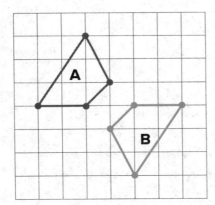

Lesson 1-9

Moves in Parallel

NAME _____ DATE _____ PERIOD _____

Learning Goal Let's transform some lines.

Warm Up
9.1 Line Moves

For each diagram, describe a translation, rotation, or reflection that takes line ℓ to line ℓ'. Then plot and label A' and B', the images of A and B.

1.

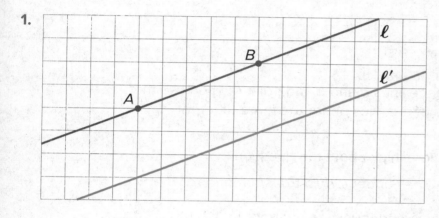

2.

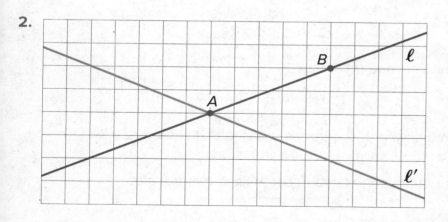

Activity

9.2 Parallel Lines

Refer to the diagram.

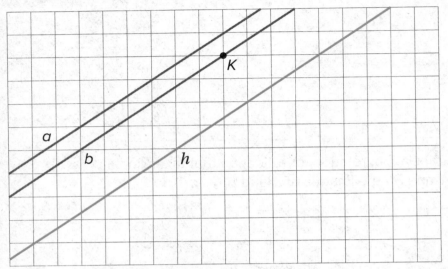

Use a piece of tracing paper to trace lines *a* and *b* and point *K*. Then use that tracing paper to draw the images of the lines under the three different transformations listed.

As you perform each transformation, think about the question:

What is the image of two parallel lines under a rigid transformation?

1. Translate lines *a* and *b* 3 units up and 2 units to the right.

   a. What do you notice about the changes that occur to lines *a* and *b* after the translation?

   b. What is the same in the original and the image?

2. Rotate lines *a* and *b* counterclockwise 180 degrees using *K* as the center of rotation.

   a. What do you notice about the changes that occur to lines *a* and *b* after the rotation?

   b. What is the same in the original and the image?

NAME _____ DATE _____ PERIOD _____

3. Reflect lines *a* and *b* across line *h*.

    a. What do you notice about the changes that occur to lines *a* and *b* after the reflection?

    b. What is the same in the original and the image?

Are you ready for more?

When you rotate two parallel lines, sometimes the two original lines intersect their images and form a quadrilateral. What is the most specific thing you can say about this quadrilateral? Can it be a square? A rhombus? A rectangle that isn't a square? Explain your reasoning.

1. The diagram shows a line with points labeled A, C, D, and B.

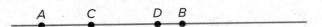

a. On the diagram, draw the image of the line and points A, C, and B after the line has been rotated 180 degrees around point D.

b. Label the images of the points A′, B′, and C′.

c. What is the order of all seven points? Explain or show your reasoning.

2. The diagram shows a line with points A and C on the line and a segment AD where D is not on the line.

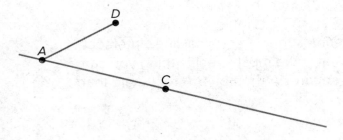

a. Rotate the figure 180 degrees about point C. Label the image of A as A′ and the image of D as D′.

b. What do you know about the relationship between angle CAD and angle CA′D′? Explain or show your reasoning.

NAME _____ DATE _____ PERIOD _____

3. The diagram shows two lines ℓ and m that intersect at a point O with point A on ℓ and point D on m.

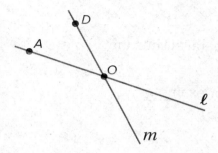

a. Rotate the figure 180 degrees around O. Label the image of A as A' and the image of D as D'.

b. What do you know about the relationship between the angles in the figure? Explain or show your reasoning.

Rigid transformations have the following properties:

1. A rigid transformation of a line is a line.

2. A rigid transformation of two parallel lines results in two parallel lines that are the same distance apart as the original two lines.

3. Sometimes, a rigid transformation takes a line to itself. For example:

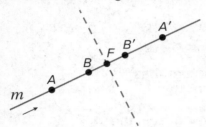

- A translation parallel to the line. The arrow shows a translation of line *m* that will take *m* to itself.

- A rotation by 180° around any point on the line. A 180° rotation of line *m* around point *F* will take *m* to itself.

- A reflection across any line perpendicular to the line. A reflection of line *m* across the dashed line will take *m* to itself.

These facts let us make an important conclusion. If two lines intersect at a point, which we'll call *O*, then a 180° rotation of the lines with center *O* shows that **vertical angles** are congruent. Here is an example:

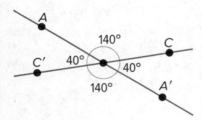

Rotating both lines by 180° around *O* sends angle *AOC* to angle *A'OC'*, proving that they have the same measure. The rotation also sends angle *AOC'* to angle *A'OC*.

Glossary

vertical angles

NAME _____ DATE _____ PERIOD _____

Practice
Moves in Parallel

1. **a.** Draw parallel lines *AB* and *CD*.

b. Pick any point *E*. Rotate *AB* 90 degrees clockwise around *E*.

c. Rotate line *CD* 90 degrees clockwise around *E*.

d. What do you notice?

2. Use the diagram to find the measures of each angle.
Explain your reasoning.

a. m∠*ABC*

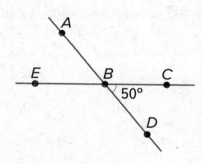

b. m∠*EBD*

c. m∠*ABE*

3. Points *P* and *Q* are plotted on a line.

   a. Find a point *R* so that a 180-degree rotation with center *R* sends *P* to *Q* and *Q* to *P*.

   b. Is there more than one point *R* that works for part a?

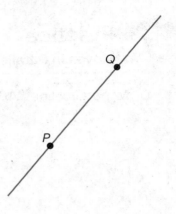

4. In the picture, triangle *A'B'C'* is an image of triangle *ABC* after a rotation. The center of rotation is *D*. **(Lesson 1-7)**

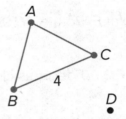
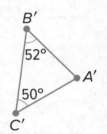

   a. What is the length of side *B'C'*? Explain how you know.

   b. What is the measure of angle *B*? Explain how you know.

   c. What is the measure of angle *C*? Explain how you know.

5. The point (-4, 1) is rotated 180 degrees counterclockwise using center (0, 0). What are the coordinates of the image? **(Lesson 1-6)**

   A. (-1, -4)

   B. (-1, 4)

   C. (4, 1)

   D. (4, -1)

Lesson 1-10

Composing Figures

NAME _____ DATE _____ PERIOD _____

Learning Goal Let's use reasoning about rigid transformations to find measurements without measuring.

Warm Up
10.1 Angles of an Isosceles Triangle

Here is a triangle.

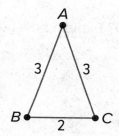

1. Reflect triangle *ABC* over line *AB*. Label the image of *C* as *C'*.

2. Rotate triangle *ABC'* around *A* so that *C'* matches up with *B*.

3. What can you say about the measures of angles *B* and *C*?

Activity

10.2 Triangle Plus One

Here is triangle *ABC*.

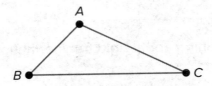

1. Draw the midpoint *M* of side *AC*.

2. Rotate triangle *ABC* 180 degrees using center *M* to form triangle *CDA*. Draw and label this triangle.

3. What kind of quadrilateral is *ABCD*? Explain how you know.

Are you ready for more?

In the activity, we made a parallelogram by taking a triangle and its image under a 180-degree rotation around the midpoint of a side.

This picture helps you justify a well-known formula for the area of a triangle.

What is the formula and how does the figure help justify it?

NAME _____ DATE _____ PERIOD _____

Activity
10.3 Triangle Plus Two

The picture shows 3 triangles. Triangle 2 and Triangle 3 are images of Triangle 1 under rigid transformations.

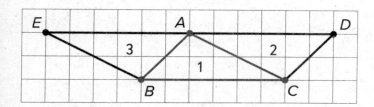

1. Describe a rigid transformation that takes Triangle 1 to Triangle 2. What points in Triangle 2 correspond to points A, B, and C in the original triangle?

2. Describe a rigid transformation that takes Triangle 1 to Triangle 3. What points in Triangle 3 correspond to points A, B, and C in the original triangle?

3. Find two pairs of line segments in the diagram that are the same length, and explain how you know they are the same length.

4. Find two pairs of angles in the diagram that have the same measure, and explain how you know they have the same measure.

Here is isosceles triangle *ONE*. Its sides *ON* and *OE* have equal lengths. Angle *O* is 30 degrees. The length of *ON* is 5 units.

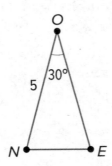

1. Reflect triangle *ONE* across segment *ON*. Label the new vertex *M*.

2. What is the measure of angle *MON*?

3. What is the measure of angle *MOE*?

4. Reflect triangle *MON* across segment *OM*. Label the point that corresponds to *N* as *T*.

5. How long is $\overline{OT}$? How do you know?

6. What is the measure of angle *TOE*?

7. If you continue to reflect each new triangle this way to make a pattern, what will the pattern look like?

NAME _____ DATE _____ PERIOD _____

Summary
Composing Figures

Earlier, we learned that if we apply a sequence of rigid transformations to a figure, then corresponding sides have equal length and corresponding angles have equal measure.

These facts let us figure out things without having to measure them!

For example, here is triangle *ABC*.

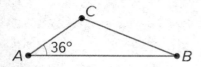

We can reflect triangle *ABC* across side *AC* to form a new triangle.

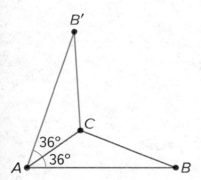

Because points *A* and *C* are on the line of reflection, they do not move. So the image of triangle *ABC* is *AB'C*.

We also know that:

• Angle *B'AC* measures 36° because it is the image of angle *BAC*.

• Segment *AB'* has the same length as segment *AB*.

When we construct figures using copies of a figure made with rigid transformations, we know that the measures of the images of segments and angles will be equal to the measures of the original segments and angles.

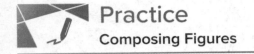
1. Here is the design for the flag of Trinidad and Tobago. Describe a sequence of translations, rotations, and reflections that take the lower left triangle to the upper right triangle.

2. Here is a picture of an older version of the flag of Great Britain. There is a rigid transformation that takes Triangle 1 to Triangle 2, another that takes Triangle 1 to Triangle 3, and another that takes Triangle 1 to Triangle 4.

   a. Measure the lengths of the sides in Triangles 1 and 2. What do you notice?

   b. What are the side lengths of Triangle 3? Explain how you know.

   c. Do all eight triangles in the flag have the same area? Explain how you know.

NAME _____ DATE _____ PERIOD _____

3. Respond to each of the following. **(Lesson 1-9)**

   a. Which of the lines in the picture is parallel to line ℓ? Explain how you know.

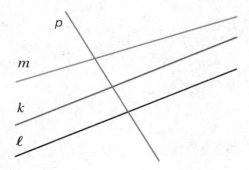

   b. Explain how to translate, rotate or reflect line ℓ to obtain line k.

   c. Explain how to translate, rotate or reflect line ℓ to obtain line p.

4. Point A has coordinates (3, 4). After a translation 4 units left, a reflection across the x-axis, and a translation 2 units down, what are the coordinates of the image? (Lesson 1-6)

5. Here is triangle XYZ:

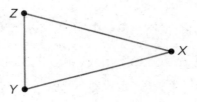

Draw these three rotations of triangle XYZ together. (Lesson 1-8)

a. Rotate triangle XYZ 90 degrees clockwise around Z.

b. Rotate triangle XYZ 180 degrees around Z.

c. Rotate triangle XYZ 270 degrees clockwise around Z.

Lesson 1-11

What Is the Same?

NAME _____ DATE _____ PERIOD _____

Learning Goal Let's decide whether shapes are the same.

Warm Up
11.1 Find the Right Hands

A person's hands are mirror images of each other. In the diagram, a left hand is labeled. Shade all of the right hands.

Activity
11.2 Are They the Same?

For each pair of shapes, decide whether or not they are the same.

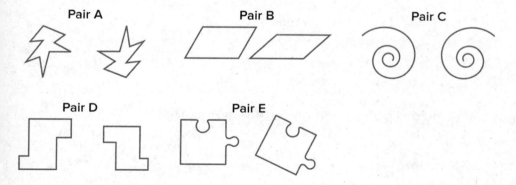

Pair A   Pair B   Pair C

Pair D   Pair E

Activity

11.3 Area, Perimeter, and Congruence

Refer to the rectangles shown.

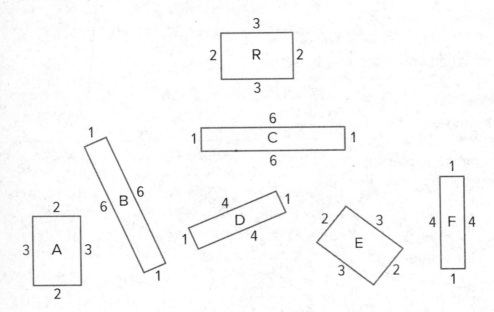

1. Which of these rectangles have the same area as Rectangle R but different perimeter?

2. Which rectangles have the same perimeter as Rectangle R but different area?

3. Which have the same area *and* the same perimeter as Rectangle R?

4. Use materials from the geometry tool kit to decide which rectangles are **congruent**. Shade congruent rectangles with the same color.

NAME _____ DATE _____ PERIOD _____

Are you ready for more?

In square *ABCD*, points *E*, *F*, *G*, and *H* are midpoints of their respective sides.

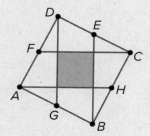

What fraction of square *ABCD* is shaded? Explain your reasoning.

Congruent is a new term for an idea we have already been using.

We say that two figures are congruent if one can be lined up exactly with the other by a sequence of rigid transformations.

For example, triangle *EFD* is congruent to triangle *ABC* because they can be matched up by reflecting triangle *ABC* across *AC* followed by the translation shown by the arrow.

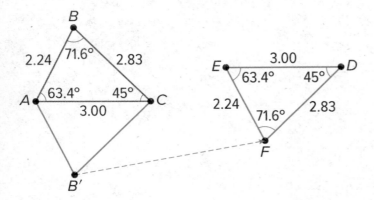

Notice that all corresponding angles and side lengths are equal.

Here are some other facts about congruent figures.

- We don't need to check all the measurements to prove two figures are congruent; we just have to find a sequence of rigid transformations that match up the figures.

- A figure that looks like a mirror image of another figure can be congruent to it. This means there must be a reflection in the sequence of transformations that matches up the figures.

- Since two congruent polygons have the same area and the same perimeter, one way to show that two polygons are *not* congruent is to show that they have a different perimeter or area.

Glossary

congruent

NAME _____ DATE _____ PERIOD _____

Practice
What Is the Same?

1. If two rectangles have the same perimeter, do they have to be congruent? Explain how you know.

2. Draw two rectangles that have the same area, but are *not* congruent.

3. For each pair of shapes, decide whether or not the two shapes are congruent. Explain your reasoning.

a.

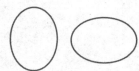

b.

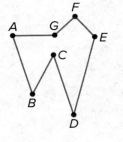

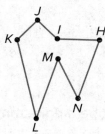

4. a. Reflect Quadrilateral A over the *x*-axis. Label the image quadrilateral B. Reflect Quadrilateral B over the *y*-axis. Label the image C.

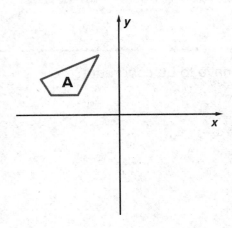

b. Are Quadrilaterals A and C congruent? Explain how you know.

5. The point (-2, -3) is rotated 90 degrees counterclockwise using center (0, 0). What are the coordinates of the image? (Lesson 1-6)

Ⓐ (-3, -2)

Ⓑ (-3, 2)

Ⓒ (3, -2)

Ⓓ (3, 2)

6. Describe a rigid transformation that takes Polygon A to Polygon B. (Lesson 1-7)

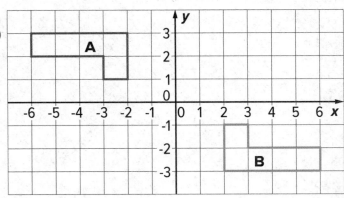

Lesson 1-12

Congruent Polygons

NAME _____ DATE _____ PERIOD _____

Learning Goal Let's decide if two figures are congruent.

Warm Up
12.1 Translated Images

All of these triangles are congruent. Sometimes we can take one figure to another with a translation. Shade the triangles that are images of triangle *ABC* under a translation.

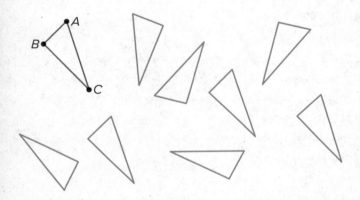

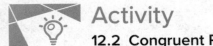
For each of the following pairs of shapes, decide whether or not they are congruent. Explain your reasoning.

1.

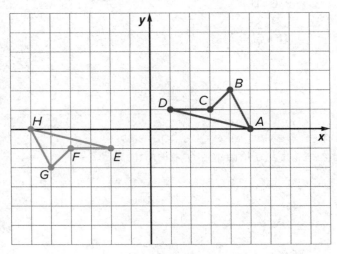

2.

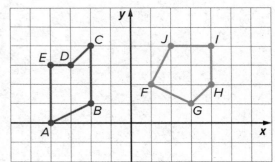

3.

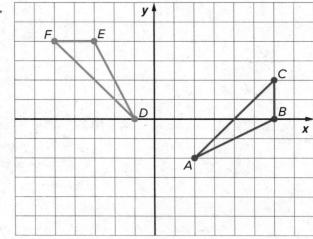

NAME _____ DATE _____ PERIOD _____

4.

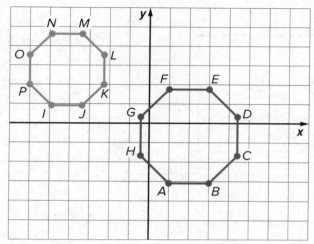

Activity

12.3 Congruent Pairs (Part 2)

For each pair of shapes, decide whether or not Shape A is congruent to
Shape B. Explain how you know.

1.

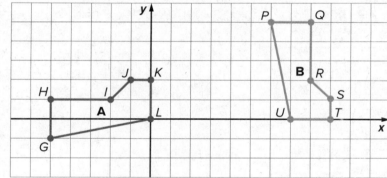

2.

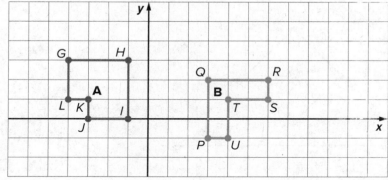

3.

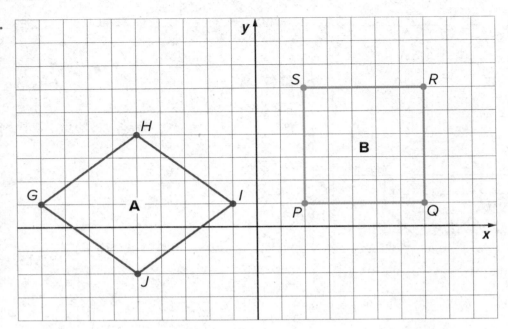

4.

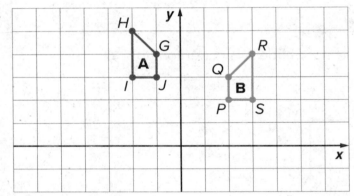

5.

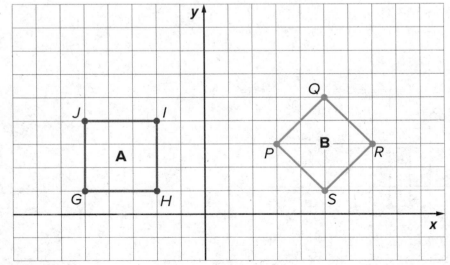

NAME _____ DATE _____ PERIOD _____

A polygon has 8 sides: five of length 1, two of length 2, and one of length 3. All sides lie on grid lines. (It may be helpful to use graph paper when working on this problem.)

1. Find a polygon with these properties.

2. Is there a second polygon, not congruent to your first, with these properties?

Activity
12.4 Building Quadrilaterals

Your teacher will give you a set of four objects.

1. Make a quadrilateral with your four objects and record what you have made.

2. Compare your quadrilateral with your partner's. Are they congruent? Explain how you know.

3. Repeat steps 1 and 2, forming different quadrilaterals. If your first quadrilaterals were not congruent, can you build a pair that is? If your first quadrilaterals were congruent, can you build a pair that is not? Explain.

How do we know if two figures are congruent?

If we copy one figure on tracing paper and move the paper so the copy covers the other figure exactly, then that suggests they are congruent.

We can prove that two figures are congruent by describing a sequence of translations, rotations, and reflections that move one figure onto the other so they match up exactly.

How do we know that two figures are *not* congruent?

If there is no correspondence between the figures where the parts have equal measure, that proves that the two figures are *not* congruent. In particular,

1. If two polygons have different sets of side lengths, they can't be congruent. For example, the figure on the left has side lengths 3, 2, 1, 1, 2, 1. The figure on the right has side lengths 3, 3, 1, 2, 2, 1. There is no way to make a correspondence between them where all corresponding sides have the same length.

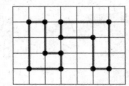

2. If two polygons have the same side lengths, but their orders can't be matched as you go around each polygon, the polygons can't be congruent. For example, rectangle *ABCD* can't be congruent to quadrilateral *EFGH*. Even though they both have two sides of length 3 and two sides of length 5, they don't correspond in the same order. In *ABCD*, the order is 3, 5, 3, 5 or 5, 3, 5, 3; in *EFGH*, the order is 3, 3, 5, 5 or 3, 5, 5, 3 or 5, 5, 3, 3.

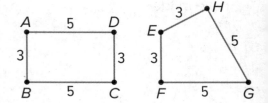

3. If two polygons have the same side lengths, in the same order, but different corresponding angles, the polygons can't be congruent. For example, parallelogram *JKLM* can't be congruent to rectangle *ABCD*. Even though they have the same side lengths in the same order, the angles are different. All angles in *ABCD* are **right angles**. In *JKLM*, angles *J* and *L* are less than 90 degrees and angles *K* and *M* are more than 90 degrees.

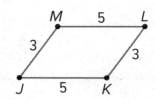

Glossary

right angle

NAME _____ DATE _____ PERIOD _____

Practice
Congruent Polygons

1. **a.** Show that the two pentagons are congruent.

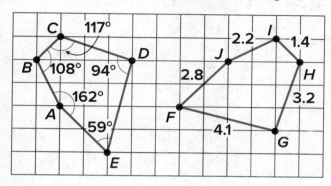

b. Find the side lengths of *ABCDE* and the angle measures of *FGHIJ*.

2. For each pair of shapes, decide whether or not the two shapes are congruent. Explain your reasoning.

a.

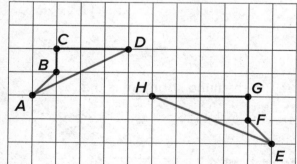

b.

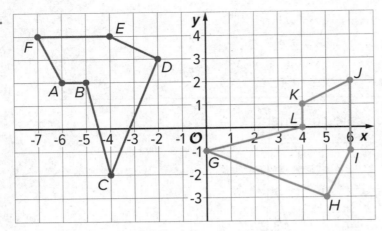

c.

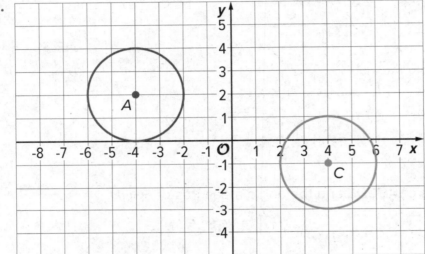

3. Respond to each question. **(Lesson 1-8)**

   a. Draw segment *PQ*.

   b. When *PQ* is rotated 180° around point *R*, the resulting segment is the same as *PQ*. Where could point *R* be located?

4. Here is trapezoid *ABCD*. Using rigid transformations on the trapezoid, build a pattern. Describe some of the rigid transformations you used. **(Lesson 1-10)**

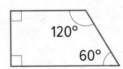

Lesson 1-13

Congruence

NAME _____ DATE _____ PERIOD _____

Learning Goal Let's find ways to test congruence of interesting figures.

Warm Up
13.1 Not Just the Vertices

Trapezoids *ABCD* and *A′B′C′D′* are congruent.

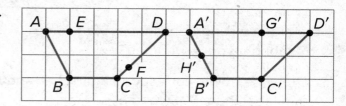

- Draw and label the points on *A′B′C′D′* that correspond to *E* and *F*.

- Draw and label the points on *ABCD* that correspond to *G′* and *H′*.

- Draw and label at least three more pairs of corresponding points.

Activity
13.2 Congruent Ovals

Are any of the ovals congruent to one another? Explain how you know.

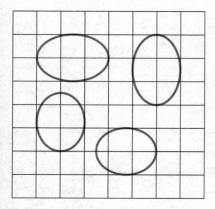

You can use 12 toothpicks to create a polygon with an area of five square toothpicks, like this:

Can you use exactly 12 toothpicks to create a polygon with an area of four square toothpicks?

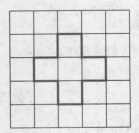

Activity
13.3 Corresponding Points in Congruent Figures

Here are two congruent shapes with some corresponding points labeled.

1. Draw the points corresponding to *B*, *D*, and *E*, and label them *B'*, *D'*, and *E'*.

2. Draw line segments *AD* and *A'D'* and measure them. Do the same for segments *BC* and *B'C'* and for segments *AE* and *A'E'*. What do you notice?

3. Do you think there could be a pair of corresponding segments with different lengths? Explain.

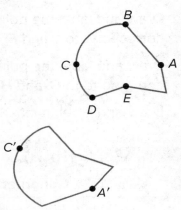

Activity
13.4 Astonished Faces

Are these faces congruent? Explain your reasoning.

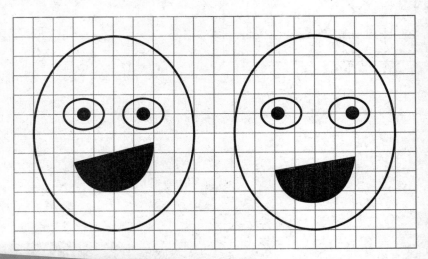

NAME _____ DATE _____ PERIOD _____

Summary
Congruence

To show two figures are congruent, you align one with the other by a sequence of rigid transformations. This is true even for figures with curved sides.

Distances between corresponding points on congruent figures are always equal, even for curved shapes. For example, corresponding segments *AB* and *A'B'* on these congruent ovals have the same length:

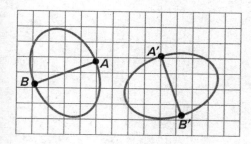

To show two figures are not congruent, you can find parts of the figures that should correspond but that have different measurements.

For example, these two ovals don't look congruent.

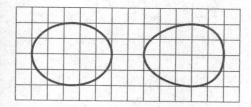

On both, the longest distance is 5 units across, and the longest distance from top to bottom is 4 units. The line segment from the highest to lowest point is in the middle of the left oval, but in the right oval, it's 2 units from the right end and 3 units from the left end. This proves they are not congruent.

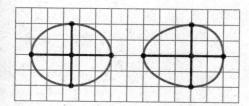

Practice
Congruence

1. Which of these four figures are congruent to the top figure?

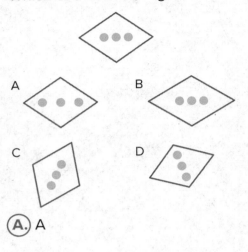

A

B

C

D

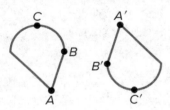

(A.) A

(B.) B

(C.) C

(D.) D

2. These two figures are congruent, with corresponding points marked.

   a. Are angles *ABC* and *A'B'C'* congruent? Explain your reasoning.

   b. Measure angles *ABC* and *A'B'C'* to check your answer.

NAME _____ DATE _____ PERIOD _____

3. Here are two figures.

Figure A   **Figure B**

Show, using measurement, that these two figures are *not* congruent.

4. Each picture shows two polygons, one labeled Polygon A and one labeled Polygon B. Describe how to move Polygon A into the position of Polygon B using a transformation. **(Lesson 1-3)**

a.

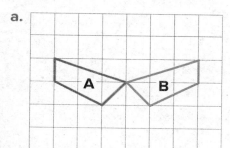

b.

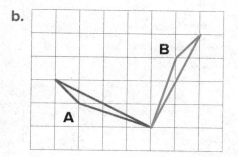

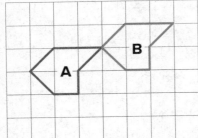

Lesson 1-14

Alternate Interior Angles

NAME _____ DATE _____ PERIOD _____

Learning Goal Let's explore why some angles are always equal.

 ## Warm Up
14.1 Angle Pairs

1. Find the measure of angle *JGH*. Explain or show your reasoning.

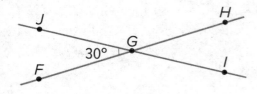

2. Find and label a second 30° degree angle in the diagram.
   Find and label an angle congruent to angle *JGH*.

 ## Activity
14.2 Cutting Parallel Lines with a Transversal

Lines *AC* and *DF* are parallel. They are cut by **transversal** *HJ*.

1. With your partner, find the seven unknown angle measures in the diagram.
   Explain your reasoning.

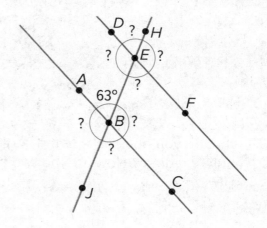

2. What do you notice about the angles with vertex *B* and the angles with vertex *E*?

3. Using what you noticed, find the measures of the four angles at point *B* in the second diagram. Lines *AC* and *DF* are parallel.

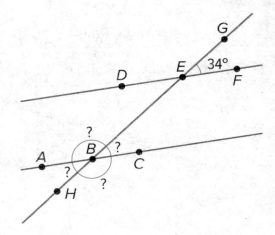

4. The next diagram resembles the first one, but the lines form slightly different angles. Work with your partner to find the six unknown angles with vertices at points *B* and *E*.

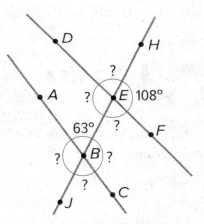

5. What do you notice about the angles in this diagram as compared to the earlier diagram? How are the two diagrams different? How are they the same?

NAME _____ DATE _____ PERIOD _____

Are you ready for more?

Parallel lines ℓ and m are cut by two transversals which intersect ℓ in the same point. Two angles are marked in the figure. Find the measure x of the third angle.

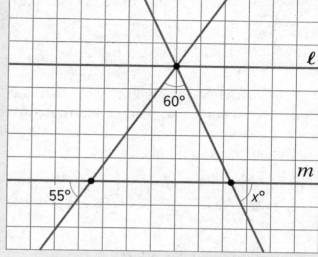

Activity

14.3 Alternate Interior Angles Are Congruent

1. Lines ℓ and k are parallel and t is a transversal. Point M is the midpoint of segment PQ.

   Find a rigid transformation showing that angles MPA and MQB are congruent.

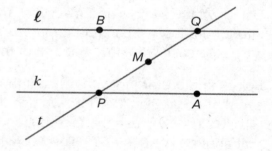

2. In this picture, lines ℓ and k are no longer parallel. M is still the midpoint of segment PQ.

   Does your argument in the earlier problem apply in this situation? Explain.

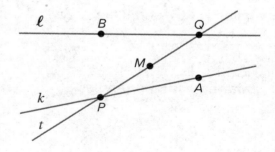

When two lines intersect, vertical angles are equal and adjacent angles are supplementary. That is, their measures sum to 180°.

For example, in this figure angles 1 and 3 are equal, angles 2 and 4 are equal, angles 1 and 4 are supplementary, and angles 2 and 3 are supplementary.

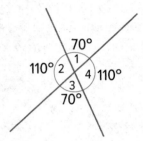

When two parallel lines are cut by another line, called a **transversal**, two pairs of **alternate interior angles** are created. ("Interior" means on the inside, or between, the two parallel lines.)

For example, in this figure angles 3 and 5 are alternate interior angles and angles 4 and 6 are also alternate interior angles.

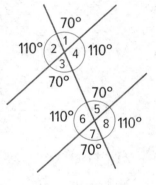

Alternate interior angles are equal because a 180° rotation around the midpoint of the segment that joins their vertices takes each angle to the other.

Imagine a point M halfway between the two intersections— can you see how rotating 180° about M takes angle 3 to angle 5?

Using what we know about vertical angles, adjacent angles, and alternate interior angles, we can find the measures of any of the eight angles created by a transversal if we know just one of them.

For example, starting with the fact that angle 1 is 70° we use vertical angles to see that angle 3 is 70°. Then we use alternate interior angles to see that angle 5 is 70°. Then we use the fact that angle 5 is supplementary to angle 8 to see that angle 8 is 110° since $180 - 70 = 110$. It turns out that there are only two different measures. In this example, angles 1, 3, 5, and 7 measure 70°, and angles 2, 4, 6, and 8 measure 110°.

Glossary

alternate interior angles
transversal

NAME _____ DATE _____ PERIOD _____

Practice
Alternate Interior Angles

1. Use the diagram to find the measure of each angle.
(Lesson 1-9)

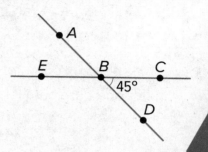

   a. m∠ABC

   b. m∠EBD

   c. m∠ABE

2. Lines k and ℓ are parallel, and the measure of angle ABC is 19 degrees.

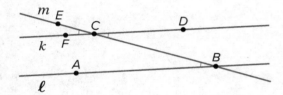

   a. Explain why the measure of angle ECF is 19 degrees. If you get stuck, consider translating line ℓ by moving B to C.

   b. What is the measure of angle BCD? Explain.

3. The diagram shows three lines with some marked angle measures. Find the missing angle measures marked with question marks.

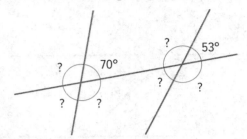

4. Lines *s* and *t* are parallel. Find the value of *x*.

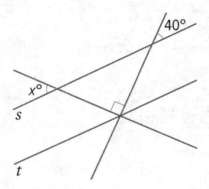

5. The two figures are scaled copies of each other.

   a. What is the scale factor that takes Figure 1 to Figure 2?

   b. What is the scale factor that takes Figure 2 to Figure 1?

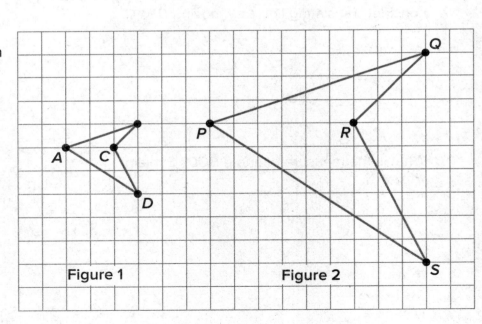

Figure 1   Figure 2

Lesson 1-15

Adding the Angles in a Triangle

NAME _____ DATE _____ PERIOD _____

Learning Goal Let's explore angles in triangles.

Warm Up
15.1 Can You Draw It?

1. Complete the table by drawing a triangle in each cell that has the properties listed for its column and row. If you think you cannot draw a triangle with those properties, write "impossible" in the cell.

2. Share your drawings with a partner. Discuss your thinking. If you disagree, work to reach an agreement.

	Acute (all angles acute)	Right (has a right angle)	Obtuse (has an obtuse angle)
Scalene (side lengths all different)			
Isosceles (at least two side lengths are equal)			
Equilateral (three side lengths equal)			

Activity
15.2 Find All Three

Your teacher will give you a card with a picture of a triangle.

1. The measurement of one of the angles is labeled. Mentally estimate the measures of the other two angles.

2. Find two other students with triangles congruent to yours but with a different angle labeled. Confirm that the triangles are congruent, that each card has a different angle labeled, and that the angle measures make sense.

3. Enter the three angle measures for your triangle on the table your teacher has posted.

Activity

15.3 Tear It Up

Your teacher will give you a page with three sets of angles and a blank space. Cut out each set of three angles. Can you make a triangle from each set that has these same three angles?

Are you ready for more?

1. Draw a quadrilateral. Cut it out, tear off its angles, and line them up. What do you notice?

2. Repeat this for several more quadrilaterals. Do you have a conjecture about the angles?

Summary

Adding the Angles in a Triangle

A 180° angle is called a **straight angle** because when it is made with two rays, they point in opposite directions and form a straight line.

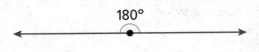

If we experiment with angles in a triangle, we find that the sum of the measures of the three angles in each triangle is 180° — the same as a straight angle!

Through experimentation we find:

- If we add the three angles of a triangle physically by cutting them off and lining up the vertices and sides, then the three angles form a straight angle.

- If we have a line and two rays that form three angles added to make a straight angle, then there is a triangle with these three angles.

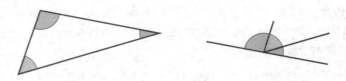

Glossary

straight angle

NAME _____ DATE _____ PERIOD _____

Practice
Adding the Angles in a Triangle

1. In triangle *ABC*, the measure of angle *A* is 40°.

   a. Give possible measures for angles *B* and *C* if triangle *ABC* is isosceles.

   b. Give possible measures for angles *B* and *C* if triangle *ABC* is right.

2. For each set of angle measures, decide if there is a triangle whose angles have these measures in degrees:

   a. 60, 60, 60

   b. 90, 90, 45

   c. 30, 40, 50

   d. 90, 45, 45

   e. 120, 30, 30

   If you get stuck, consider making a line segment. Then use a protractor to measure angles with the first two angle measures.

3. Angle *A* in triangle *ABC* is obtuse. Can angle *B* or angle *C* be obtuse? Explain your reasoning.

4. For each pair of polygons, describe the transformation that could be applied to Polygon A to get Polygon B. (Lesson 1-3)

a.

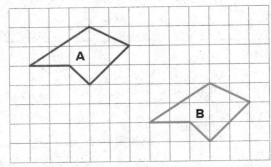

b.

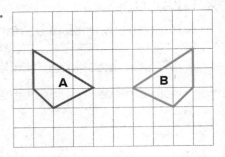

c.

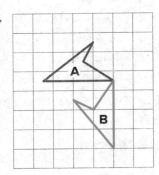

5. On the grid, draw a scaled copy of quadrilateral *ABCD* using a scale factor of $\frac{1}{2}$. (Lesson 1-14)

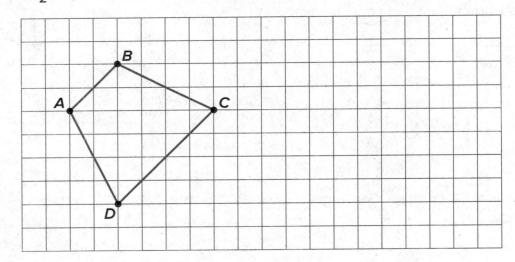

Lesson 1-16

Parallel Lines and the Angles in a Triangle

NAME _____ DATE _____ PERIOD _____

Learning Goal Let's see why the angles in a triangle add to 180 degrees.

Warm Up

16.1 True or False: Computational Relationships

Is each equation true or false?

1. $62 - 28 = 60 - 30$

2. $3 \cdot \text{-}8 = (2 \cdot \text{-}8) - 8$

3. $\frac{16}{\text{-}2} + \frac{24}{\text{-}2} = \frac{40}{\text{-}2}$

Activity

16.2 Angle Plus Two

Here is triangle *ABC*.

1. Rotate triangle *ABC* 180° around the midpoint of side *AC*. Label the new vertex *D*.

2. Rotate triangle *ABC* 180° around the midpoint of side *AB*. Label the new vertex *E*.

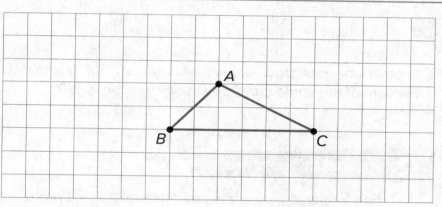

3. Look at angles *EAB*, *BAC*, and *CAD*. Without measuring, write what you think is the sum of the measures of these angles. Explain or show your reasoning.

4. Is the measure of angle *EAB* equal to the measure of any angle in triangle *ABC*? If so, which one? If not, how do you know?

5. Is the measure of angle *CAD* equal to the measure of any angle in triangle *ABC*? If so, which one? If not, how do you know?

6. What is the sum of the measures of angles *ABC*, *BAC*, and *ACB*?

Activity

16.3 Every Triangle in the World

Here is △*ABC*. Line *DE* is parallel to line *AC*.

1. What is $m\angle DBA + b + m\angle CBE$? Explain how you know.

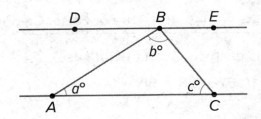

2. Use your answer to explain why $a + b + c = 180$.

3. Explain why your argument will work for *any* triangle: that is, explain why the sum of the angle measures in *any* triangle is 180°.

Are you ready for more?

1. Using a ruler, create a few quadrilaterals. Use a protractor to measure the four angles inside the quadrilateral. What is the sum of these four angle measures?

2. Come up with an explanation for why anything you notice must be true (hint: draw one diagonal in each quadrilateral).

NAME _____ DATE _____ PERIOD _____

Activity

16.4 Four Triangles Revisited

This diagram shows a square *BDFH* that has been made by images of triangle *ABC* under rigid transformations.

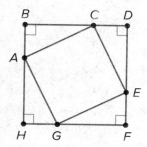

Given that angle *BAC* measures 53 degrees, find as many other angle measures as you can.

Summary

Parallel Lines and the Angles in a Triangle

Using parallel lines and rotations, we can understand why the angles in a triangle always add to 180°.

Here is triangle *ABC*. Line *DE* is parallel to *AC* and contains *B*.

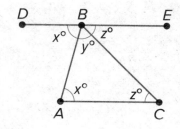

- A 180-degree rotation of triangle *ABC* around the midpoint of *AB* interchanges angles *A* and *DBA* so they have the same measure. In the picture these angles are marked as $x°$.

- A 180-degree rotation of triangle *ABC* around the midpoint of *BC* interchanges angles *C* and *CBE* so they have the same measure. In the picture, these angles are marked as $z°$.

Also, *DBE* is a straight line because 180-degree rotations take lines to parallel lines. So, the three angles with vertex *B* make a line and they add up to 180° ($x + y + z = 180$).

But x, y, z are the measures of the three angles in $\triangle ABC$ so the sum of the angles in a triangle is always 180°!

1. For each triangle, find the measure of the missing angle.

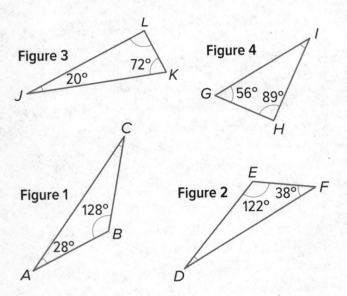

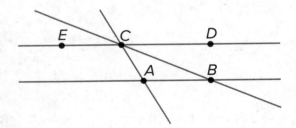

2. Is there a triangle with *two* right angles? Explain your reasoning.

3. In this diagram, lines *AB* and *CD* are parallel.

Angle *ABC* measures 35° and angle *BAC* measures 115°.

   a. What is $m\angle ACE$?

   b. What is $m\angle DCB$?

   c. What is $m\angle ACB$?

NAME _____ DATE _____ PERIOD _____

4. Here is a diagram of triangle *DEF*.

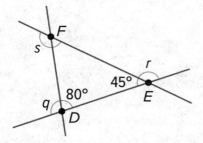

a. Find the measures of angles *q*, *r*, and *s*.

b. Find the sum of the measures of angles *q*, *r*, and *s*.

c. What do you notice about these three angles?

5. The two figures are congruent. (Lesson 1-13)

a. Label the points A′, B′ and C′ that correspond to A, B, and C in the figure on the right.

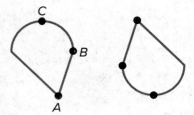

b. If segment AB measures 2 cm, how long is segment A′B′? Explain.

c. The point D is shown in addition to A and C. How can you find the point D′ that corresponds to D? Explain your reasoning.

Lesson 1-17

Rotate and Tessellate

NAME _____ DATE _____ PERIOD _____

Learning Goal Let's make complex patterns using transformations.

Warm Up
17.1 Deducing Angle Measures

Your teacher will give you some shapes.

1. How many copies of the equilateral triangle can you fit together around a single vertex, so that the triangles' edges have no gaps or overlaps? What is the measure of each angle in these triangles?

2. What are the measures of the angles in the...

   a. square?

   b. hexagon?

   c. parallelogram?

   d. right triangle?

   e. octagon?

   f. pentagon?

Activity

17.2 Tessellate This

1. Design your own **tessellation**. You will need to decide which shapes you want to use and make copies. Remember that a tessellation is a repeating pattern that goes on forever to fill up the entire plane.

2. Find a partner and trade pictures. Describe a transformation of your partner's picture that takes the pattern to itself. How many different transformations can you find that take the pattern to itself? Consider translations, reflections, and rotations.

3. If there's time, color and decorate your tessellation.

Activity

17.3 Rotate That

1. Make a design with rotational symmetry.

2. Find a partner who has also made a design. Exchange designs and find a transformation of your partner's design that takes it to itself. Consider rotations, reflections, and translations.

3. If there's time, color and decorate your design.

Glossary
_____

tessellation

Learning Targets

Lesson	Learning Target(s)
1-1 Moving in the Plane	• I can describe how a figure moves and turns to get from one position to another.
1-2 Naming the Moves	• I can identify corresponding points before and after a transformation. • I know the difference between translations, rotations, and reflections.
1-3 Grid Moves	• I can decide which type of transformations will work to move one figure to another. • I can use grids to carry out transformations of figures.
1-4 Making the Moves	• I can use the terms translation, rotation, and reflection to precisely describe transformations.

(continued on the next page)

(continued from the previous page)

Lesson	Learning Target(s)
1-5 Coordinate Moves	• I can apply transformations to points on a grid if I know their coordinates.
1-6 Describing Transformations	• I can apply transformations to a polygon on a grid if I know the coordinates of its vertices.
1-7 No Bending or Stretching	• I can describe the effects of a rigid transformation on the lengths and angles in a polygon.
1-8 Rotation Patterns	• I can describe how to move one part of a figure to another using a rigid transformation.

Lesson	Learning Target(s)
1-9 Moves in Parallel	• I can describe the effects of a rigid transformation on a pair of parallel lines.
	• If I have a pair of vertical angles and know the angle measure of one of them, I can find the angle measure of the other.
1-10 Composing Figures	• I can find missing side lengths or angle measures using properties of rigid transformations.
1-11 What Is the Same?	• I can decide visually whether or not two figures are congruent.
1-12 Congruent Polygons	• I can decide using rigid transformations whether or not two figures are congruent.
1-13 Congruence	• I can use distances between points to decide if two figures are congruent.

(continued on the next page)

(continued from the previous page)

Lesson	Learning Target(s)
1-14 Alternate Interior Angles	• If I have two parallel lines cut by a transversal, I can identify alternate interior angles and use that to find missing angle measurements.
1-15 Adding the Angles in a Triangle	• If I know two of the angle measures in a triangle, I can find the third angle measure.
1-16 Parallel Lines and the Angles in a Triangle	• I can explain using pictures why the sum of the angles in any triangle is 180 degrees.
1-17 Rotate and Tessellate	• I can repeatedly use rigid transformations to make interesting repeating patterns of figures. • I can use properties of angle sums to reason about how figures will fit together.

Unit 2

Dilations, Similarity, and Introducing Slope

At the end of this unit, you'll apply what you learned about similarity to examine the length of shadows of different objects.

Topics
- Dilations
- Similarity
- Slope
- Let's Put It to Work

Dilations, Similarity, and Introducing Slope

Lesson 2-1

Projecting and Scaling

NAME _____ DATE _____ PERIOD _____

Learning Goal Let's explore scaling.

Warm Up
1.1 Number Talk: Remembering Fraction Division

Find each quotient. Write your answer as a fraction or a mixed number.

1. $6\frac{1}{4} \div 2$ **2.** $10\frac{1}{7} \div 5$ **3.** $8\frac{1}{2} \div 11$

Activity
1.2 Sorting Rectangles

Rectangles were made by cutting an $8\frac{1}{2}$-inch by 11-inch piece of paper in half, in half again, and so on, as illustrated in the diagram. Find the lengths of each rectangle and enter them in the appropriate table.

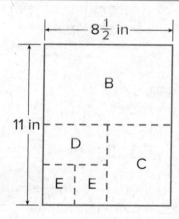

1. Some of the rectangles are scaled copies of the full sheet of paper (Rectangle A). Record the measurements of those rectangles in this table.

Rectangle	Length of Short Side (inches)	Length of Long Side (inches)
A	$8\frac{1}{2}$	11

2. Some of the rectangles are *not* scaled copies of the full sheet of paper. Record the measurements of those rectangles in this table.

Rectangle	Length of Short Side (inches)	Length of Long Side (inches)

3. Look at the measurements for the rectangles that are scaled copies of the full sheet of paper. What do you notice about the measurements of these rectangles? Look at the measurements for the rectangles that are *not* scaled copies of the full sheet. What do you notice about these measurements?

4. Stack the rectangles that are scaled copies of the full sheet so that they all line up at a corner, as shown in the diagram. Do the same with the other set of rectangles. On each stack, draw a line from the bottom left corner to the top right corner of the biggest rectangle. What do you notice?

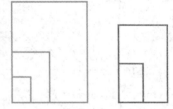

5. Stack *all* of the rectangles from largest to smallest so that they all line up at a corner. Compare the lines that you drew. Can you tell, from the drawn lines, which set each rectangle came from?

Are you ready for more?

In many countries, the standard paper size is not 8.5 inches by 11 inches (called "letter" size), but instead 210 millimeters by 297 millimeters (called "A4" size). Are these two rectangle sizes scaled copies of one another?

NAME _____ DATE _____ PERIOD _____

Activity

1.3 Scaled Rectangles

Here is a picture of Rectangle R, which has been evenly divided into smaller rectangles. Two of the smaller rectangles are labeled B and C.

Rectangle R

1. Is B a scaled copy of R? If so, what is the **scale factor**?

2. Is C a scaled copy of B? If so, what is the scale factor?

3. Is C a scaled copy of R? If so, what is the scale factor?

Summary

Projecting and Scaling

Scaled copies of rectangles have an interesting property. Can you see what it is?

Here, the larger rectangle is a scaled copy of the smaller one (with a scale factor of $\frac{3}{2}$). Notice how the diagonal of the large rectangle contains the diagonal of the smaller rectangle. This is the case for any two scaled copies of a rectangle if we line them up as shown. If two rectangles are *not* scaled copies of one another, then the diagonals do not match up. In this unit, we will investigate how to make scaled copies of a figure.

Glossary

scale factor

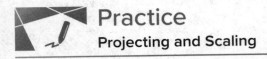

Practice

Projecting and Scaling

1. Rectangle *A* measures 12 cm by 3 cm. Rectangle *B* is a scaled copy of Rectangle *A*. Select **all** of the measurement pairs that could be the dimensions of Rectangle *B*.

(A.) 6 cm by 1.5 cm

(B.) 10 cm by 2 cm

(C.) 13 cm by 4 cm

(D.) 18 cm by 4.5 cm

(E.) 80 cm by 20 cm

2. Rectangle *A* has length 12 and width 8. Rectangle *B* has length 15 and width 10. Rectangle *C* has length 30 and width 15.

   a. Is Rectangle *A* a scaled copy of Rectangle *B*? If so, what is the scale factor?

   b. Is Rectangle *B* a scaled copy of Rectangle *A*? If so, what is the scale factor?

   c. Explain how you know that Rectangle *C* is *not* a scaled copy of Rectangle *B*.

   d. Is Rectangle *A* a scaled copy of Rectangle *C*? If so, what is the scale factor?

NAME _____ DATE _____ PERIOD _____

3. Here are three polygons.

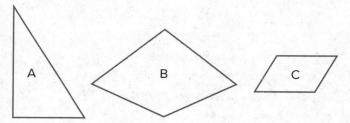

a. Draw a scaled copy of Polygon A with scale factor $\frac{1}{2}$.

b. Draw a scaled copy of Polygon B with scale factor 2.

c. Draw a scaled copy of Polygon C with scale factor $\frac{1}{4}$.

4. Which of these sets of angle measures could be the three angles in a triangle? (Lesson 1-15)

(A.) 40°, 50°, 60°

(B.) 50°, 60°, 70°

(C.) 60°, 70°, 80°

(D.) 70°, 80°, 90°

5. In the picture, lines *AB* and *CD* are parallel. Find the measures of the following angles. Explain your reasoning. (Lesson 1-14)

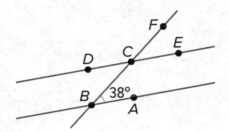

a. ∠BCD

b. ∠ECF

c. ∠DCF

Lesson 2-2

Circular Grid

NAME _____ DATE _____ PERIOD _____

Learning Goal Let's dilate figures on circular grids.

Warm Up
2.1 Notice and Wonder: Concentric Circles

What do you notice? What do you wonder?

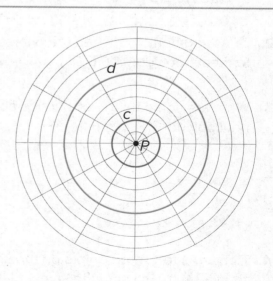

Activity
2.2 A Droplet on the Surface

The larger Circle *d* is a **dilation** of the smaller Circle *c*. *P* is the **center of dilation**.

1. Draw four points *on* the smaller circle (not inside the circle!), and label them *E*, *F*, *G*, and *H*.

2. Draw the rays from *P* through each of those four points.

3. Label the points where the rays meet the larger circle *E′*, *F′*, *G′*, and *H′*.

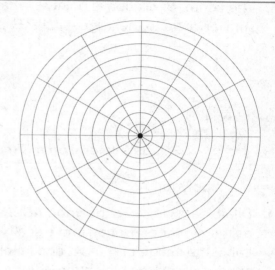

4. Complete the table. In the row labeled *c*, write the distance between *P* and the point on the smaller circle in grid units. In the row labeled *d*, write the distance between *P* and the corresponding point on the larger circle in grid units.

	E	F	G	H
c				
d				

5. The center of dilation is point *P*. What is the *scale factor* that takes the smaller circle to the larger circle? Explain your reasoning.

Activity

2.3 Quadrilateral on a Circular Grid

Here is a polygon *ABCD*.

1. Dilate each vertex of polygon *ABCD* using *P* as the center of dilation and a scale factor of 2. Label the image of *A* as *A′*, and label the images of the remaining three vertices as *B′*, *C′*, and *D′*.

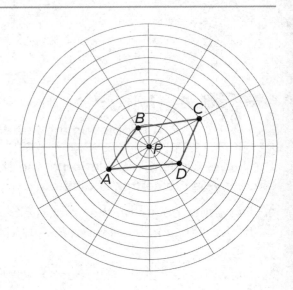

2. Draw segments between the dilated points to create polygon *A′B′C′D′*.

3. What are some things you notice about the new polygon?

4. Choose a few more points on the sides of the original polygon and transform them using the same dilation. What do you notice?

NAME _____ DATE _____ PERIOD _____

5. Dilate each vertex of polygon *ABCD* using *P* as the center of dilation and a scale factor of $\frac{1}{2}$. Label the image of *A* as *E*, the image of *B* as *F*, the image of *C* as *G* and the image of *D* as *H*.

6. What do you notice about polygon *EFGH*?

Are you ready for more?

Suppose *P* is a point not on line segment $\overline{WX}$. Let $\overline{YZ}$ be the dilation of line segment $\overline{WX}$ using *P* as the center with scale factor 2. Experiment using a circular grid to make predictions about whether each of the following statements must be true, might be true, or must be false.

1. $\overline{YZ}$ is twice as long $\overline{WX}$.

2. $\overline{YZ}$ is five units longer than $\overline{WX}$.

3. The point *P* is on $\overline{YZ}$.

4. $\overline{YZ}$ and $\overline{WX}$ intersect.

Activity

2.4 A Quadrilateral and Concentric Circles

Dilate polygon *EFGH* using *Q* as the center of dilation and a scale factor of $\frac{1}{3}$.

The image of *F* is already shown on the diagram. (You may need to draw more rays from *Q* in order to find the images of other points.)

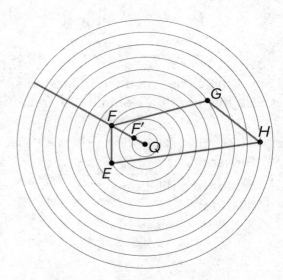

A circular grid like this one can be helpful for performing **dilations**.

The radius of the smallest circle is one unit, and the radius of each successive circle is one unit more than the previous one.

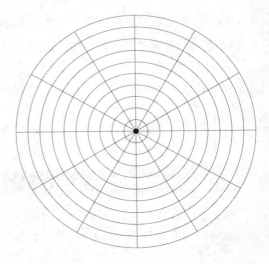

To perform a dilation, we need a **center of dilation**, a scale factor, and a point to dilate. In the picture, *P* is the center of dilation. With a scale factor of 2, each point stays on the same ray from *P*, but its distance from *P* doubles:

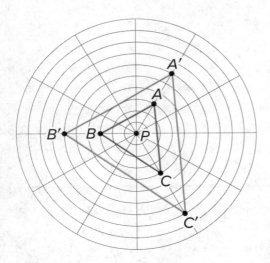

Since the circles on the grid are the same distance apart, segment *PA′* has twice the length of segment *PA*, and the same holds for the other points.

Glossary

center of a dilation
dilation

NAME _____ DATE _____ PERIOD _____

Practice
Circular Grid

1. Here are Circles *c* and *d*. Point *O* is the center of dilation, and the dilation takes Circle *c* to Circle *d*.

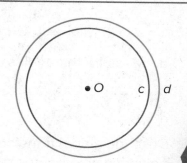

   a. Plot a point on Circle *c*. Label the point *P*. Plot where *P* goes when the dilation is applied.

   b. Plot a point on Circle *d*. Label the point *Q*. Plot a point that the dilation takes to *Q*.

2. Here is triangle *ABC*.

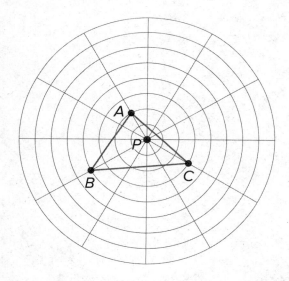

   a. Dilate each vertex of triangle *ABC* using *P* as the center of dilation and a scale factor of 2. Draw the triangle connecting the three new points.

   b. Dilate each vertex of triangle *ABC* using *P* as the center of dilation and a scale factor of $\frac{1}{2}$. Draw the triangle connecting the three new points.

c. Measure the longest side of each of the three triangles. What do you notice?

d. Measure the angles of each triangle. What do you notice?

3. Describe a rigid transformation that you could use to show the polygons are congruent. **(Lesson 1-12)**

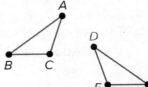

4. The line has been partitioned into three angles. Is there a triangle with these three angle measures? Explain. **(Lesson 1-15)**

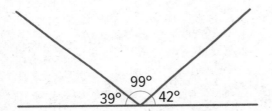

Lesson 2-3

Dilations with No Grid

NAME _____ DATE _____ PERIOD _____

Learning Goal Let's dilate figures not on grids.

Warm Up
3.1 Points on a Ray

1. Find and label a point *C* on the ray whose distance from *A* is twice the distance from *B* to *A*.

2. Find and label a point *D* on the ray whose distance from *A* is half the distance from *B* to *A*.

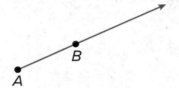

Activity
3.2 Dilation Obstacle Course

Here is a diagram that shows nine points.

1. Dilate *B* using a scale factor of 5 and *A* as the center of dilation. Which point is its image?

2. Using *H* as the center of dilation, dilate *G* so that its image is *E*. What scale factor did you use?

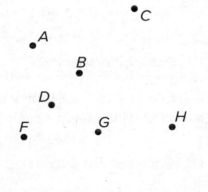

3. Using *H* as the center of dilation, dilate *E* so that its image is *G*. What scale factor did you use?

4. To dilate *F* so that its image is *B*, what point on the diagram can you use as a center?

5. Dilate *H* using *A* as the center and a scale factor of $\frac{1}{3}$. Which point is its image?

6. Describe a dilation that uses a labeled point as its center and that would take *F* to *H*.

7. Using *B* as the center of dilation, dilate *H* so that its image is itself. What scale factor did you use?

Activity

3.3 Getting Perspective

1. Using one colored pencil, draw the images of points *P* and *Q* below using *C* as the center of dilation and a scale factor of 4. Label the new points *P'* and *Q'*.

2. Using a different color, draw the images of points *P* and *Q* using *C* as the center of dilation and a scale factor of $\frac{1}{2}$. Label the new points *P"* and *Q"*.

Pause here so your teacher can review your diagram. Your teacher will then give you a scale factor to use in the next part.

NAME _____ DATE _____ PERIOD _____

3. Now you'll make a perspective drawing. Here is a rectangle.

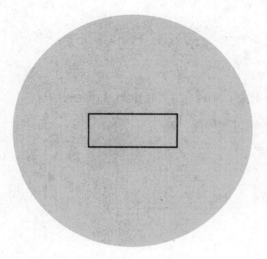

a. Choose a point *inside the shaded circular region* but *outside the rectangle* to use as the center of dilation. Label it *C*.

b. Using your center *C* and the scale factor you were given, draw the image under the dilation of each vertex of the rectangle, one at a time. Connect the dilated vertices to create the dilated rectangle.

c. Draw a segment that connects each of the original vertices with its image. This will make your diagram look like a cool three-dimensional drawing of a box! If there's time, you can shade the sides of the box to make it look more realistic.

d. Compare your drawing to other people's drawings. What is the same and what is different? How do the choices you made affect the final drawing? Was your dilated rectangle closer to *C* than to the original rectangle, or farther away? How is that decided?

Here is line segment *DE* and its image *D'E'* under a dilation.

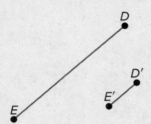

1. Use a ruler to find and draw the center of dilation. Label it *F*.

2. What is the scale factor of the dilation?

Summary
Dilations with No Grid

If *A* is the center of dilation, how can we find which point is the dilation of *B* with scale factor 2?

Since the scale factor is larger than 1, the point must be farther away from *A* than *B* is, which makes *C* the point we are looking for. If we measure the distance between *A* and *C*, we would find that it is exactly twice the distance between *A* and *B*.

A dilation with scale factor less than 1 brings points closer. The point *D* is the dilation of *B* with center *A* and scale factor $\frac{1}{3}$.

NAME _____ DATE _____ PERIOD _____

Practice
Dilations with No Grid

1. Segment *AB* measures 3 cm. Point *O* is the center of dilation. How long is the image of *AB* after a dilation with . . .

   a. scale factor 5?

   b. scale factor 3.7?

   c. scale factor $\frac{1}{5}$?

   d. scale factor *s*?

2. Here are points *A* and *B*. Plot the points for each dilation described.

   a. *C* is the image of *B* using *A* as the center of dilation and a scale factor of 2.

   b. *D* is the image of *A* using *B* as the center of dilation and a scale factor of 2.

   c. *E* is the image of *B* using *A* as the center of dilation and a scale factor of $\frac{1}{2}$.

   d. *F* is the image of *A* using *B* as the center of dilation and a scale factor of $\frac{1}{2}$.

3. Make a perspective drawing. Include in your work the center of dilation, the shape you dilate, and the scale factor you use.

4. Triangle *ABC* is a scaled copy of triangle *DEF*. Side *AB* measures 12 cm and is the longest side of *ABC*. Side *DE* measures 8 cm and is the longest side of *DEF*. **(Lesson 2-1)**

   a. Triangle *ABC* is a scaled copy of triangle *DEF* with what scale factor?

   b. Triangle *DEF* is a scaled copy of triangle *ABC* with what scale factor?

5. The diagram shows two intersecting lines. Find the missing angle measures. **(Lesson 1-14)**

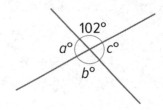

6. Respond to each of the following.
   (Lesson 1-12)

   a. Show that the two triangles are congruent.

   b. Find the side lengths of *DEF* and the angle measures of *ABC*.

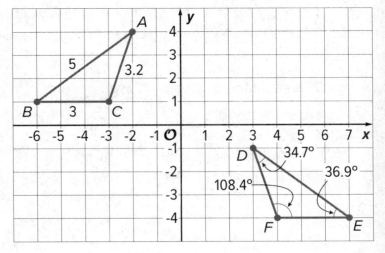

Lesson 2-4

Dilations on a Square Grid

NAME _____ DATE _____ PERIOD _____

Learning Goal Let's dilate figures on a square grid.

Warm Up
4.1 Estimating a Scale Factor

Point *C* is the dilation of point *B* with center of dilation *A* and scale factor *s*. Estimate *s*. Be prepared to explain your reasoning.

Activity
4.2 Dilations on a Grid

1. Find the dilation of quadrilateral *ABCD* with center *P* and scale factor 2.

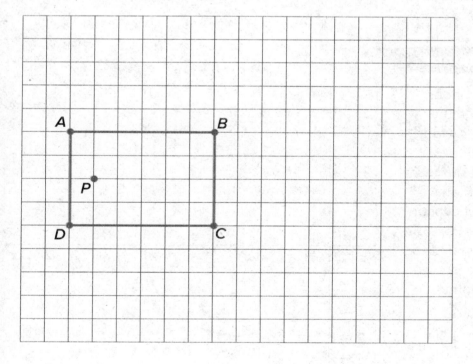

2. Find the dilation of triangle *QRS* with center *T* and scale factor 2.

3. Find the dilation of triangle *QRS* with center *T* and scale factor $\frac{1}{2}$.

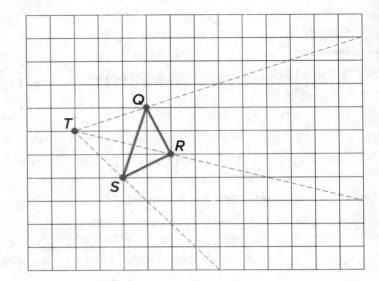

 ## Activity

4.3 Card Sort: Matching Dilations on a Coordinate Grid

Your teacher will give you some cards. Each of Cards 1 through 6 shows a figure in the coordinate plane and describes a dilation.

Each of Cards A through E describes the image of the dilation for one of the numbered cards.

Match number cards with letter cards. One of the number cards will not have a match. For this card, you'll need to draw an image.

Are you ready for more?

The image of a circle under dilation is a circle when the center of the dilation is the center of the circle. What happens if the center of dilation is a point on the circle? Using center of dilation (0, 0) and scale factor 1.5, dilate the circle shown on the diagram. This diagram shows some points to try dilating.

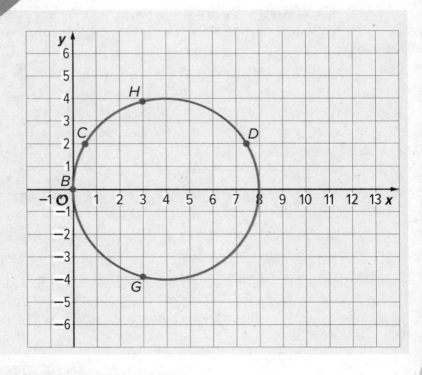

NAME _____ DATE _____ PERIOD _____

Summary
Dilations on a Square Grid

Square grids can be useful for showing dilations. The grid is helpful especially when the center of dilation and the point(s) being dilated lie at grid points. Rather than using a ruler to measure the distance between the points, we can count grid units.

For example, suppose we want to dilate point Q with center of dilation P and scale factor $\frac{3}{2}$. Since Q is 4 grid squares to the left and 2 grid squares down from P, the dilation will be 6 grid squares to the left and 3 grid squares down from P (can you see why?). The dilated image is marked as Q' in the picture.

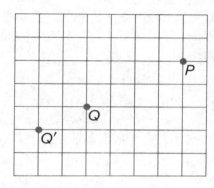

Sometimes the square grid comes with coordinates. The coordinate grid gives us a convenient way to *name* points, and sometimes the coordinates of the image can be found with just arithmetic.

For example, to make a dilation with center (0, 0) and scale factor 2 of the triangle with coordinates (-1, -2), (3, 1), and (2, -1), we can just double the coordinates to get (-2, -4), (6, 2), and (4, -2).

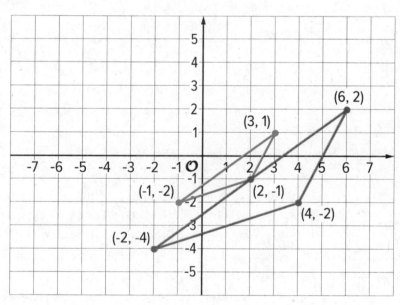

1. Triangle *ABC* is dilated using *D* as the center of dilation with scale factor 2. The image is triangle *A'B'C'*. Clare says the two triangles are congruent, because their angle measures are the same. Do you agree? Explain how you know.

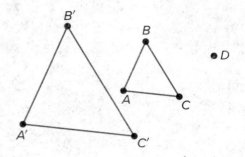

2. Sketch the image of quadrilateral *PQRS* under the following dilations.

   a. The dilation centered at *R* with scale factor 2.

   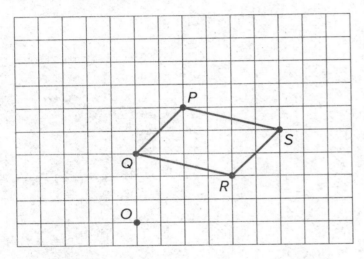

   b. The dilation centered at *O* with scale factor $\frac{1}{2}$.

   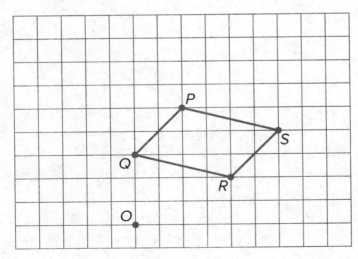

NAME _____ DATE _____ PERIOD _____

c. The dilation centered at *S* with scale factor $\frac{1}{2}$.

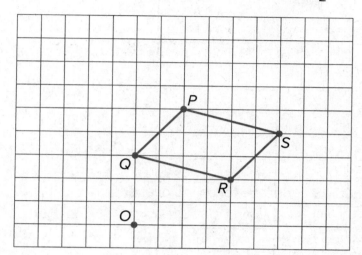

3. The diagram shows three lines with some marked angle measures. Find the missing angle measures marked with question marks. **(Lesson 1-14)**

4. Describe a sequence of translations, rotations, and reflections that takes Polygon P to Polygon Q. (Lesson 1-4)

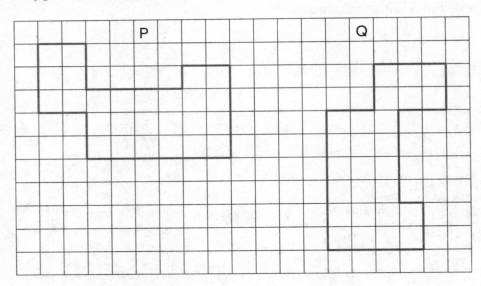

5. Point *B* has coordinates (-2, -5). After a translation 4 units down, a reflection across the *y*-axis, and a translation 6 units up, what are the coordinates of the image? (Lesson 1-6)

Lesson 2-5

More Dilations

NAME _____ DATE _____ PERIOD _____

Learning Goal Let's look at dilations in the coordinate plane.

Warm Up
5.1 Many Dilations of a Triangle

All of the triangles are dilations of Triangle D. The dilations use the same center
P, but different scale factors.

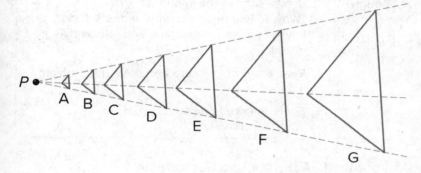

1. What do Triangles A, B, and C have in common?

2. What do Triangles E, F, and G have in common?

3. What does this tell us about the different scale factors used?

Activity
5.2 Info Gap: Dilations

Your teacher will give you either a *problem card* or a *data card*. Do not show or read your card to your partner.

If your teacher gives you the problem card:	If your teacher gives you the data card:
1. Silently read your card and think about what information you need to be able to answer the question. 2. Ask your partner for the specific information that you need. 3. Explain how you are using the information to solve the problem. Continue to ask questions until you have enough information to solve the problem. 4. Share the *problem card* and solve the problem independently. 5. Read the *data card* and discuss your reasoning.	1. Silently read your card. 2. Ask your partner *"What specific information do you need?"* and wait for them to *ask* for information. If your partner asks for information that is not on the card, do not do the calculations for them. Tell them you don't have that information. 3. Before sharing the information, ask *"Why do you need that information?"* Listen to your partner's reasoning and ask clarifying questions. 4. Read the *problem card* and solve the problem independently. 5. Share the *data card* and discuss your reasoning.

Pause here so your teacher can review your work. Ask your teacher for a new set of cards and repeat the activity, trading roles with your partner.

NAME _____ DATE _____ PERIOD _____

Are you ready for more?

Triangle *EFG* was created by dilating triangle *ABC* using a scale factor of 2 and center *D*. Triangle *HIJ* was created by dilating triangle *ABC* using a scale factor of $\frac{1}{2}$ and center *D*.

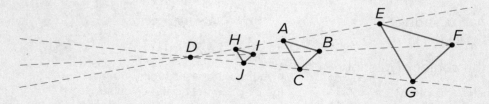

1. What would the image of triangle *ABC* look like under a dilation with scale factor 0?

2. What would the image of the triangle look like under dilation with a scale factor of -1? If possible, draw it and label the vertices *A'*, *B'*, and *C'*. If it's not possible, explain why not.

3. If possible, describe what happens to a shape if it is dilated with a negative scale factor. If dilating with a negative scale factor is not possible, explain why not.

Summary
More Dilations

One important use of coordinates is to communicate geometric information precisely.

Let's consider a quadrilateral *ABCD* in the coordinate plane. Performing a dilation of *ABCD* requires three vital pieces of information:

1. the coordinates of *A*, *B*, *C*, and *D*

2. the coordinates of the center of dilation, *P*

3. the scale factor of the dilation

With this information, we can dilate the vertices *A*, *B*, *C*, and *D* and then draw the corresponding segments to find the dilation of *ABCD*.

Without coordinates, describing the location of the new points would likely require sharing a picture of the polygon and the center of dilation.

NAME _____ DATE _____ PERIOD _____

Practice
More Dilations

1. Quadrilateral *ABCD* is dilated with center (0, 0), taking *B* to *B'*. Draw *A'B'C'D'*.

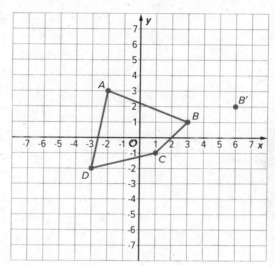

2. Triangles B and C have been built by dilating Triangle A.

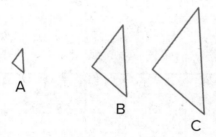

a. Find the center of dilation.

b. Triangle *B* is a dilation of *A* with approximately what scale factor?

c. Triangle *A* is a dilation of *B* with approximately what scale factor?

d. Triangle *B* is a dilation of *C* with approximately what scale factor?

3. Here is a triangle.

   a. Draw the dilation of triangle *ABC*, with center (0, 0), and scale factor 2. Label this triangle *A'B'C'*.

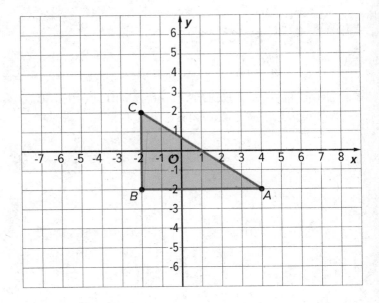

   b. Draw the dilation of triangle *ABC*, with center (0, 0), and scale factor $\frac{1}{2}$. Label this triangle *A"B"C"*.

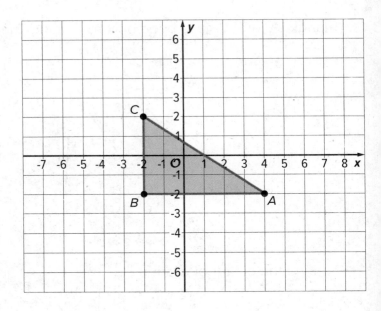

   c. Is *A"B"C"* a dilation of triangle *A'B'C'*? If yes, what are the center of dilation and the scale factor?

4. Triangle *DEF* is a right triangle, and the measure of angle *D* is 28°. What are the measures of the other two angles? **(Lesson 1-15)**

Lesson 2-6

Similarity

NAME _____ DATE _____ PERIOD _____

Learning Goal Let's explore similar figures.

Warm Up
6.1 Equivalent Expressions

Use what you know about operations and their properties to write three expressions equivalent to the expression shown.

$10(2 + 3) - 8 \cdot 3$

Activity
6.2 Similarity Transformations (Part 1)

1. Triangle *EGH* and triangle *LME* are **similar**. Find a sequence of translations, rotations, reflections, and dilations that shows this.

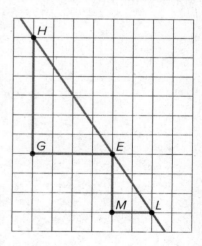

2. Hexagon *ABCDEF* and hexagon *HGLKJI* are similar. Find a sequence of translations, rotations, reflections, and dilations that shows this.

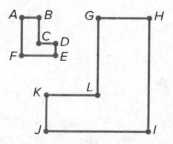

The same sequence of transformations takes Triangle A to Triangle B, takes Triangle B to Triangle C, and so on. Describe a sequence of transformations with this property.

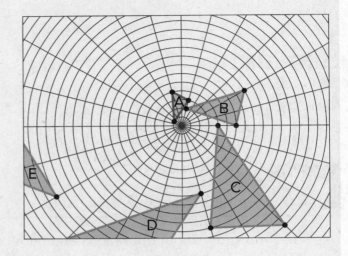

Activity

6.3 Similarity Transformations (Part 2)

Sketch figures similar to Figure A that use only the transformations listed to show similarity.

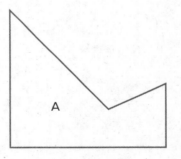

1. A translation and a reflection. Label your sketch Figure B.
   Pause here so your teacher can review your work.

NAME _____ DATE _____ PERIOD _____

2. A reflection and a dilation with scale factor greater than 1. Label your sketch Figure C.

3. A rotation and a reflection. Label your sketch Figure D.

4. A dilation with scale factor less than 1 and a translation. Label your sketch Figure E.

Activity

6.4 Methods for Translations and Dilations

Your teacher will give you a set of five cards and your partner a different set of five cards. Using only the cards you were given, find at least one way to show that triangle *ABC* and triangle *DEF* are similar.

Compare your method with your partner's method. What is the same about your methods? What is different?

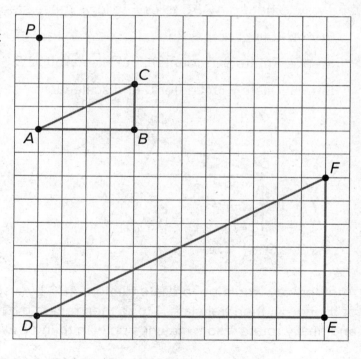

Let's show that triangle *ABC* is similar to triangle *DEF*:

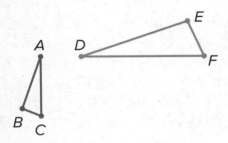

Two figures are **similar** if one figure can be transformed into the other by a sequence of translations, rotations, reflections, and dilations.

There are many correct sequences of transformations, but we only need to describe one to show that two figures are similar.

One way to get from *ABC* to *DEF* follows these steps:

*   Step 1: Reflect across line *f*.

*   Step 2: Rotate 90° counterclockwise around *D*.

*   Step 3: Dilate with center *D* and scale factor 2.

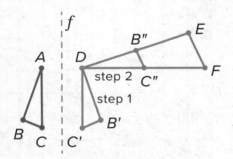

Another way would be to dilate triangle *ABC* by a scale factor of 2 with center of dilation *A*, then translate *A* to *D*, then reflect over a vertical line through *D*, and finally rotate it so it matches up with triangle *DEF*.

What steps would you choose to show the two triangles are similar?

Glossary

similar

NAME _____ DATE _____ PERIOD _____

Practice
Similarity

1. Each diagram has a pair of figures, one larger than the other. For each pair, show that the two figures are similar by identifying a sequence of translations, rotations, reflections, and dilations that takes the smaller figure to the larger one.

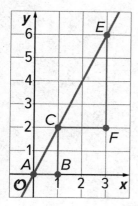

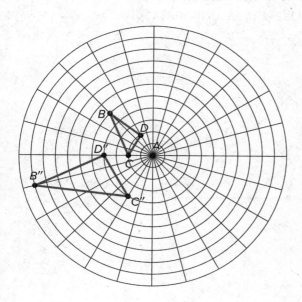

2. Here are two similar polygons. Measure the side lengths and angles of each polygon. What do you notice?

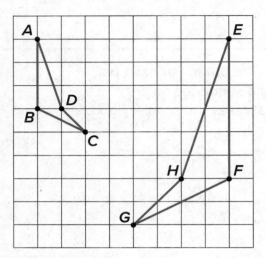

3. Each figure shows a pair of similar triangles, one contained in the other. For each pair, describe a point and a scale factor to use for a dilation moving the larger triangle to the smaller one. Use a measurement tool to find the scale factor.

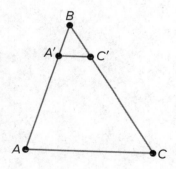

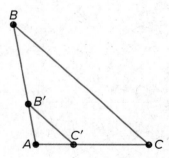

Lesson 2-7

Similar Polygons

NAME _____ DATE _____ PERIOD _____

Learning Goal Let's look at sides and angles of similar polygons.

 ## Warm Up
7.1 All, Some, None: Congruence and Similarity

Choose whether each of the statements is true in *all* cases, in *some* cases, or in *no* cases.

1. If two figures are congruent, then they are similar.

2. If two figures are similar, then they are congruent.

3. If an angle is dilated with the center of dilation at its vertex, the angle measure may change.

 ## Activity
7.2 Are They Similar?

1. Let's look at a square and a rhombus.

   Priya says, "These polygons are similar because their side lengths are all the same." Clare says, "These polygons are not similar because the angles are different." Do you agree with either Priya or Clare? Explain your reasoning.

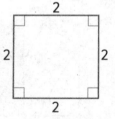
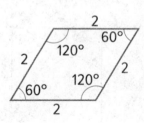

2. Now, let's look at rectangles *ABCD* and *EFGH*.

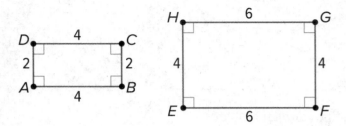

Jada says, "These rectangles are similar because all of the side lengths differ by 2."

Lin says, "These rectangles are similar. I can dilate *AD* and *BC* using a scale factor of 2 and *AB* and *CD* using a scale factor of 1.5 to make the rectangles congruent. Then I can use a translation to line up the rectangles."
Do you agree with either Jada or Lin? Explain your reasoning.

Are you ready for more?

Points *A* through *H* are translated to the right to create points *A'* through *H'*.

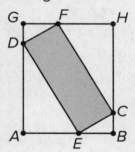 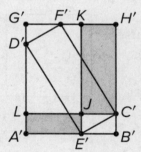

All of the following are rectangles: *GHBA*, *FCED*, *KH'C'J*, and *LJE'A'*. Which is greater, the area of blue rectangle *DFCE* or the total area of yellow rectangles *KH'C'J* and *LJE'A'*?

NAME _____ DATE _____ PERIOD _____

Activity

7.3 Find Someone Similar

Your teacher will give you a card. Find someone else in the room who has a card with a polygon that is similar but not congruent to yours. When you have found your partner, work with them to explain how you know that the two polygons are similar.

Are you ready for more?

On the left is an equilateral triangle where dashed lines have been added, showing how you can partition an equilateral triangle into smaller similar triangles.

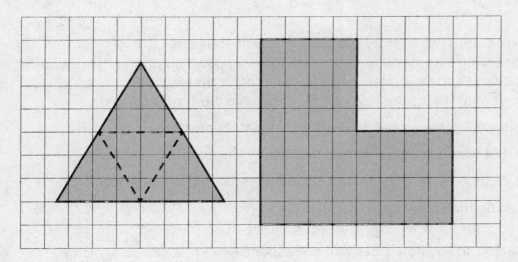

Find a way to do this for the figure on the right, partitioning it into smaller figures which are each similar to that original shape. What's the fewest number of pieces you can use? The most?

When two polygons are similar:

- Every angle and side in one polygon has a corresponding part in the other polygon.

- All pairs of corresponding angles have the same measure.

- Corresponding sides are related by a single scale factor. Each side length in one figure is multiplied by the scale factor to get the corresponding side length in the other figure.

Consider the two rectangles shown here. Are they similar?

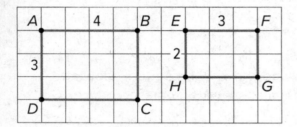

It looks like rectangles *ABCD* and *EFGH* could be similar, if you match the long edges and match the short edges. All the corresponding angles are congruent because they are all right angles. Calculating the scale factor between the sides is where we see that "looks like" isn't enough to make them similar.

To scale the long side *AB* to the long side *EF*, the scale factor must be $\frac{3}{4}$, because $4 \cdot \frac{3}{4} = 3$. But the scale factor to match *AD* to *EH* has to be $\frac{2}{3}$, because $3 \cdot \frac{2}{3} = 2$. So, the rectangles are not similar because the scale factors for all the parts must be the same.

Here is an example that shows how sides can correspond (with a scale factor of 1), but the quadrilaterals are not similar because the angles don't have the same measure:

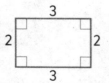 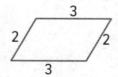

NAME _____ DATE _____ PERIOD _____

Practice
Similar Polygons

1. Triangle *DEF* is a dilation of triangle *ABC* with scale factor 2. In triangle *ABC*, the largest angle measures 82°. What is the largest angle measure in triangle *DEF*?

   (A.) 41°   (C.) 123°

   (B.) 82°   (D.) 164°

2. Draw two polygons that are similar but could be mistaken for not being similar. Explain why they are similar.

3. Draw two polygons that are *not* similar but could be mistaken for being similar. Explain why they are not similar.

4. These two triangles are similar. Find side lengths *a* and *b*. *Note:* The two figures are not drawn to scale.

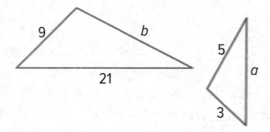

5. Jada claims that $B'C'D'$ is a dilation of BCD using A as the center of dilation. What are some ways you can convince Jada that her claim is not true? (Lesson 2-3)

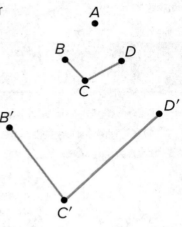

6. Respond to each of the following. (Lesson 1-8)

  a. Draw a horizontal line segment AB.

  b. Rotate segment AB 90° counterclockwise around point A. Label any new points.

  c. Rotate segment AB 90° clockwise around point B. Label any new points.

  d. Describe a transformation on segment AB you could use to finish building a square.

Lesson 2-8

Similar Triangles

NAME _____ DATE _____ PERIOD _____

Learning Goal Let's look at similar triangles.

Warm Up
8.1 Equivalent Expressions

Create three different expressions that are each equal to 20. Each expression should include only these three numbers: 4, -2, and 10.

Activity
8.2 Making Pasta Angles and Triangles

Your teacher will give you some dried pasta and a set of angles.

1. Create a triangle using three pieces of pasta and angle *A*. Your triangle *must* include the angle you were given, but you are otherwise free to make any triangle you like. Tape your pasta triangle to a sheet of paper so it won't move.

   a. After you have created your triangle, measure each side length with a ruler and record the length on the paper next to the side. Then measure the angles to the nearest 5 degrees using a protractor and record these measurements on your paper.

   b. Find two others in the room who have the same angle *A* and compare your triangles. What is the same? What is different? Are the triangles congruent? Similar?

   c. How did you decide if they were or were not congruent or similar?

2. Now use more pasta and angles *A*, *B*, and *C* to create another triangle. Tape this pasta triangle on a separate sheet of paper.

   a. After you have created your triangle, measure each side length with a ruler and record the length on the paper next to the side. Then measure the angles to the nearest 5 degrees using a protractor and record these measurements on your paper.

   b. Find two others in the room who used your same angles and compare your triangles. What is the same? What is different? Are the triangles congruent? Similar?

   c. How did you decide if they were or were not congruent or similar?

3. Here is triangle *PQR*. Break a new piece of pasta, different in length than segment *PQ*.

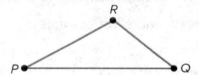

   • Tape the piece of pasta so that it lays on top of line *PQ* with one end of the pasta at *P* (if it does not fit on the page, break it further). Label the other end of the piece of pasta *S*.

   • Tape a full piece of pasta, with one end at *S*, making an angle congruent to ∠*PQR*.

   • Tape a full piece of pasta on top of line *PR* with one end of the pasta at *P*. Call the point where the two full pieces of pasta meet *T*.

   a. Is your new pasta triangle *PST* similar to △*PQR*? Explain your reasoning.

   b. If your broken piece of pasta were a different length, would the pasta triangle still be similar to △*PQR*? Explain your reasoning.

NAME _____ DATE _____ PERIOD _____

Quadrilaterals *ABCD* and *EFGH* have four angles measuring 240°, 40°, 40°, and 40°. Do *ABCD* and *EFGH* have to be similar?

Activity

8.3 Similar Figures in a Regular Pentagon

1. This diagram has several triangles that are similar to triangle *DJI*.

   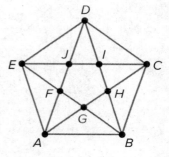

   a. Three different scale factors were used to make triangles similar to *DJI*. In the diagram, find at least one triangle of each size that is similar to *DJI*.

   b. Explain how you know each of these three triangles is similar to *DJI*.

2. Find a triangle in the diagram that is not similar to *DJI*.

Figure out how to draw some more lines in the pentagon diagram to make more triangles similar to *DJI*.

We learned earlier that two polygons are similar when there is a sequence of translations, rotations, reflections, and dilations taking one polygon to the other.

When the polygons are triangles, we only need to check that both triangles have two corresponding angles to show they are similar—can you tell why?

Here is an example. Triangle *ABC* and triangle *DEF* each have a 30-degree angle and a 45-degree angle.

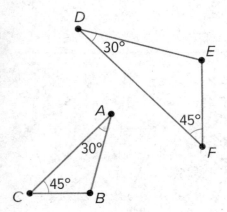

We can translate *A* to *D* and then rotate so that the two 30-degree angles are aligned, giving this picture:

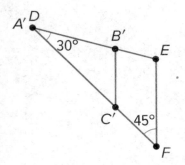

Now a dilation with center *D* and appropriate scale factor will move *C'* to *F*. This dilation also moves *B'* to *E*, showing that triangles *ABC* and *DEF* are similar.

NAME _____ DATE _____ PERIOD _____

Practice
Similar Triangles

1. In each pair, some of the angles of two triangles in degrees are given.
Use the information to decide if the triangles are similar or not.
Explain how you know.

- Triangle A: 53, 71, ____; Triangle B: 53, 71, ____

- Triangle C: 90, 37, ____; Triangle D: 90, 53, ____

- Triangle E: 63, 45, _____; Triangle F: 14, 71, _____

- Triangle G: 121, ____, ____; Triangle H: 70, ____, ____

2. a. Draw two equilateral triangles that are not congruent.

b. Measure the side lengths and angles of your triangles. Are the two triangles similar?

c. Do you think two equilateral triangles will be similar *always*, *sometimes*, or *never*? Explain your reasoning.

3. In the figure, line *BC* is parallel to line *DE*. Explain why △*ABC* is similar to △*ADE*.

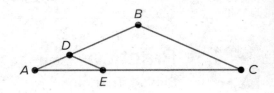

4. The quadrilateral *PQRS* in the diagram is a parallelogram. Let *P'Q'R'S'* be the image of *PQRS* after applying a dilation centered at a point O (not shown) with scale factor 3. (Lesson 2-4)

   Which of the following is true?

   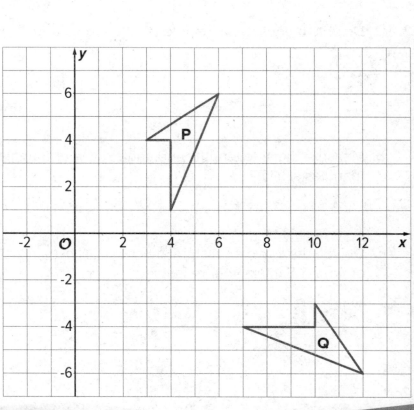

   (A.) $P'Q' = PQ$

   (B.) $P'Q' = 3PQ$

   (C.) $PQ = 3P'Q'$

   (D.) cannot be determined from the information given

5. Describe a sequence of transformations for which Quadrilateral P is the image of Quadrilateral Q. (Lesson 1-6)

Lesson 2-9

Side Length Quotients in Similar Triangles

NAME _____ DATE _____ PERIOD _____

Learning Goal Let's find missing side lengths in triangles.

Warm Up
9.1 Two-three-four and Four-five-six

Triangle A has side lengths 2, 3, and 4. Triangle B has side lengths 4, 5, and 6. Is Triangle A similar to Triangle B?

Activity
9.2 Quotients of Sides Within Similar Triangles

Triangle *ABC* is similar to triangles *DEF*, *GHI*, and *JKL*. The scale factors for the dilations that show triangle *ABC* is similar to each triangle are in the table.

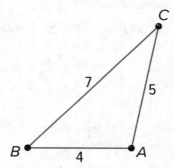

1. Find the side lengths of triangles *DEF*, *GHI*, and *JKL*. Record them in the table.

Triangle	Scale Factor	Length of Short Side	Length of Medium Side	Length of Long Side
ABC	1	4	5	7
DEF	2			
GHI	3			
JKL	$\frac{1}{2}$			

2. Your teacher will asign you one of the three columns. For all four triangles, find the quotient of the triangle side lengths assigned to you and record them in the table. What do you notice about the quotients?

Triangle	(Long Side) ÷ (Short Side)	(Long Side) ÷ (Medium Side)	(Medium Side) ÷ (Short Side)
ABC	$\frac{7}{4}$ or 1.75		
DEF			
GHI			
JKL			

3. Compare your results with your partner's and complete your table.

Are you ready for more?

Triangles *ABC* and *DEF* are similar. Explain why $\frac{AB}{BC} = \frac{DE}{EF}$.

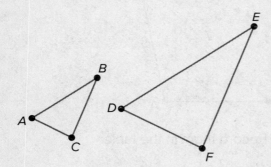

NAME _____ DATE _____ PERIOD _____

Activity
9.3 Using Side Quotients to Find Side Lengths of Similar Triangles

Triangles *ABC*, *EFD*, and *GHI* are all similar. The side lengths of the triangles all have the same units. Find the unknown side lengths.

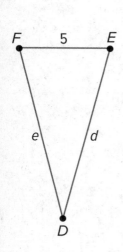

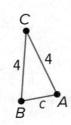

If two polygons are similar, then the side lengths in one polygon are multiplied by the same scale factor to give the corresponding side lengths in the other polygon. For these triangles the scale factor is 2:

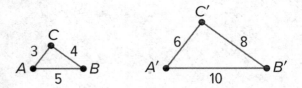

Here is a table that shows relationships between the short and medium length sides of the small and large triangle.

	Small Triangle	Large Triangle
Medium Side	4	8
Short Side	3	6
(Medium Side) ÷ (Short Side)	$\frac{4}{3}$	$\frac{8}{6} = \frac{4}{3}$

The lengths of the medium side and the short side are in a ratio of 4 : 3. This means that the medium side in each triangle is $\frac{4}{3}$ as long as the short side.

This is true for all similar polygons; the ratio between two sides in one polygon is the same as the ratio of the corresponding sides in a similar polygon.

We can use these facts to calculate missing lengths in similar polygons. For example, triangles $A'B'C'$ and ABC shown here are similar. Let's find the length of segment $B'C'$.

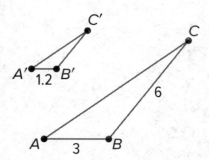

In triangle ABC, side BC is twice as long as side AB, so this must be true for any triangle that is similar to triangle ABC. Since $A'B'$ is 1.2 units long and $2 \cdot 1.2 = 2.4$, the length of side $B'C'$ is 2.4 units.

NAME _____ DATE _____ PERIOD _____

Practice
Side Length Quotients in Similar Triangles

1. These two triangles are similar. What are *a* and *b*?
   Note: the two figures are not drawn to scale.

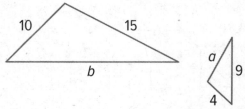

2. Here is triangle *ABC*. Triangle *XYZ* is similar to *ABC*
   with scale factor $\frac{1}{4}$.

   a. Draw what triangle *XYZ* might look like.

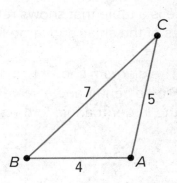

   b. How do the angle measures of triangle *XYZ* compare to triangle *ABC*?
      Explain how you know.

   c. What are the side lengths of triangle *XYZ*?

   d. For triangle *XYZ*, calculate (long side) ÷ (medium side), and compare to
      triangle *ABC*.

3. The two triangles shown are similar. Find the value of $\frac{d}{c}$.

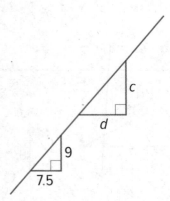

4. The diagram shows two nested triangles that share a vertex. Find a center and a scale factor for a dilation that would move the larger triangle to the smaller triangle. **(Lesson 2-5)**

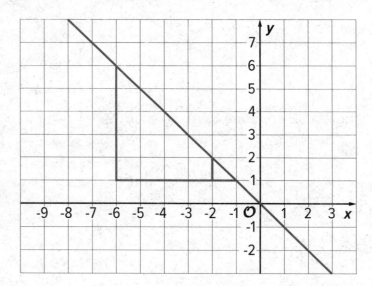

Lesson 2-10

Meet Slope

NAME _____ DATE _____ PERIOD _____

Learning Goal Let's learn about the slope of a line.

Warm Up
10.1 Equal Quotients

Write some numbers that are equal to 15 ÷ 12.

Activity
10.2 Similar Triangles on the Same Line

1. The figure shows three right triangles, each with its longest side on the same line.

   Your teacher will assign you two triangles. Explain why the two triangles are similar.

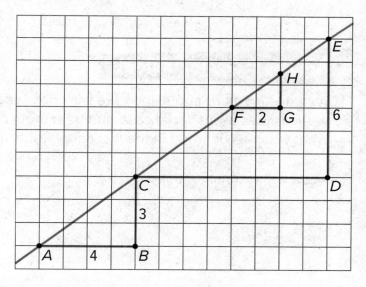

2. Complete the table.

Triangle	Length of Vertical Side	Length of Horizontal Side	(Vertical Side) ÷ (Horizontal Side)
ABC			
CDE			
FGH			

Activity

10.3 Multiple Lines with the Same Slope

1. Draw two lines with slope 3. What do you notice about the two lines?

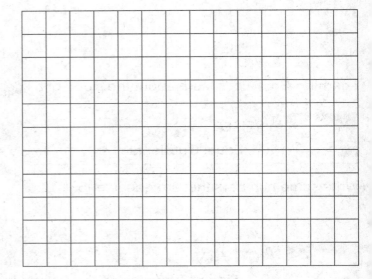

2. Draw two lines with slope $\frac{1}{2}$. What do you notice about the two lines?

Are you ready for more?

As we learn more about lines, we will occasionally have to consider perfectly vertical lines as a special case and treat them differently. Think about applying what you have learned in the last couple of activities to the case of vertical lines.

What is the same? What is different?

NAME _____ DATE _____ PERIOD _____

Activity
10.4 Different Slopes of Different Lines

Here are several lines.

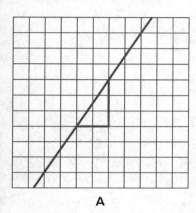

A

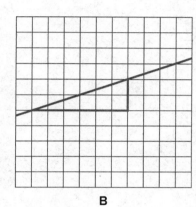

B

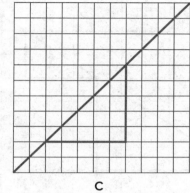

C

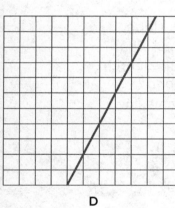

D

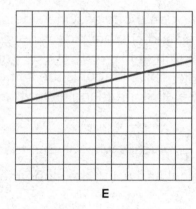

E

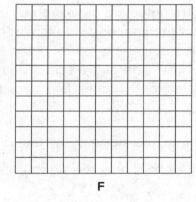

F

1. Match each line shown with a slope from this list: $\frac{1}{3}$, 2, 1, 0.25, $\frac{3}{2}$, $\frac{1}{2}$.

2. One of the given slopes does not have a line to match. Draw a line with this slope on the empty grid (F).

Here is a line drawn on a grid. There are also four right triangles drawn. Do you notice anything the triangles have in common?

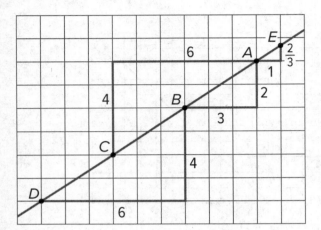

These four triangles are all examples of *slope triangles*. One side of a slope triangle is on the line, one side is vertical, and another side is horizontal. The **slope** of the line is the quotient of the length of the vertical side and the length of the horizontal side of the slope triangle. This number is the same for *all* slope triangles for the same line because all slope triangles for the same line are similar.

In this example, the slope of the line is $\frac{2}{3}$, which is what all four triangles have in common. Here is how the slope is calculated using the slope triangles:

- Points A and B give $2 \div 3 = \frac{2}{3}$

- Points D and B give $4 \div 6 = \frac{2}{3}$

- Points A and C give $4 \div 6 = \frac{2}{3}$

- Points A and E give $\frac{2}{3} \div 1 = \frac{2}{3}$

Glossary

slope

NAME _____ DATE _____ PERIOD _____

Practice
Meet Slope

1. Of the three lines in the graph, one has slope 1, one has slope 2, and one has slope $\frac{1}{5}$. Label each line with its slope.

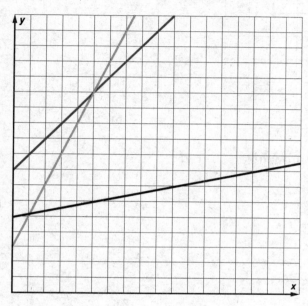

2. Draw three lines with slope 2, and three lines with slope $\frac{1}{3}$. What do you notice?

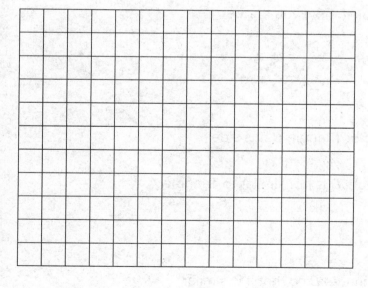

3. The figure shows two right triangles, each with its longest side on the same line.

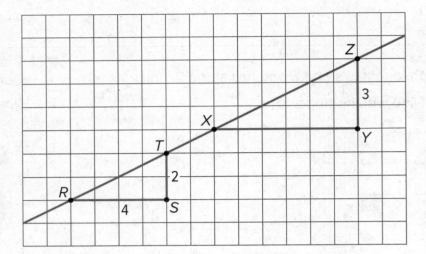

a. Explain how you know the two triangles are similar.

b. How long is *XY*?

c. For each triangle, calculate (vertical side) ÷ (horizontal side).

d. What is the slope of the line? Explain how you know.

4. Triangle *A* has side lengths 3, 4, and 5. Triangle *B* has side lengths 6, 7, and 8. (Lesson 2-9)

a. Explain how you know that Triangle *B* is *not* similar to Triangle *A*.

b. Give possible side lengths for Triangle *B* so that it is similar to Triangle *A*.

Lesson 2-11

Writing Equations for Lines

NAME _____ DATE _____ PERIOD _____

Learning Goal Let's explore the relationship between points on a line and the slope of the line.

 ## Warm Up
11.1 Coordinates and Lengths in the Coordinate Plane

Find each of the following and explain your reasoning.

1. The length of segment *BE*

2. The coordinates of *E*

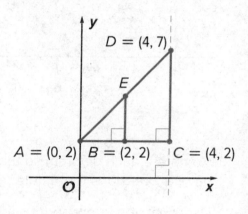

 ## Activity
11.2 What We Mean by an Equation of a Line

Line *j* is shown in the coordinate plane.

1. What are the coordinates of *B* and *D*?

2. Is point (20, 15) on line *j*? Explain how you know.

3. Is point (100, 75) on line *j*? Explain how you know.

4. Is point (90, 68) on line *j*? Explain how you know.

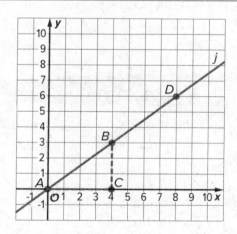

5. Suppose you know the *x*- and *y*-coordinates of a point. Write a rule that would allow you to test whether the point is on line *j*.

Activity

11.3 Writing Relationships from Slope Triangles

Here are two diagrams.

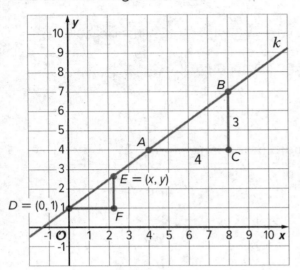 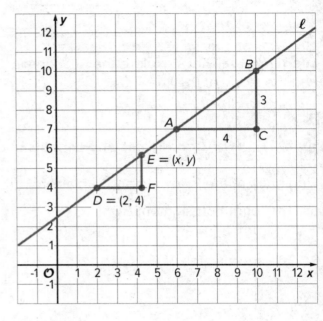

1. Complete each diagram so that all vertical and horizontal segments have expressions for their lengths.

2. Use what you know about similar triangles to find an equation for the quotient of the vertical and horizontal side lengths of △DFE in each diagram.

Are you ready for more?

1. Find the area of the shaded region by summing the areas of the shaded triangles.

2. Find the area of the shaded region by subtracting the area of the unshaded region from the large triangle.

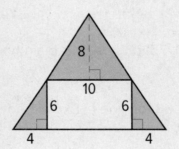

3. What is going on here?

NAME _____ DATE _____ PERIOD _____

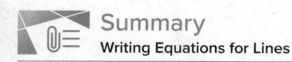

Summary
Writing Equations for Lines

Here are the points *A*, *C*, and *E* on the same line. Triangles *ABC* and *ADE* are slope triangles for the line so we know they are similar triangles.

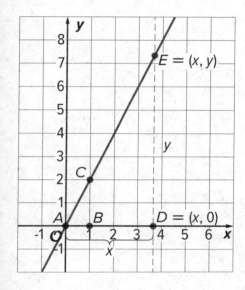

Let's use their similarity to better understand the relationship between *x* and *y*, which make up the coordinates of point *E*.

The slope for triangle *ABC* is $\frac{2}{1}$ since the vertical side has length 2 and the horizontal side has length 1.

The slope we find for triangle *ADE* is $\frac{y}{x}$ because the vertical side has length *y* and the horizontal side has length *x*.

These two slopes must be equal since they are from slope triangles for the same line, and so: $\frac{2}{1} = \frac{y}{x}$.

Since $\frac{2}{1} = 2$ this means that the value of *y* is twice the value of *x*, or that *y* = 2*x*. This equation is true for any point (*x*, *y*) on the line!

Here are two different slope triangles. We can use the same reasoning to describe the relationship between x and y for this point E.

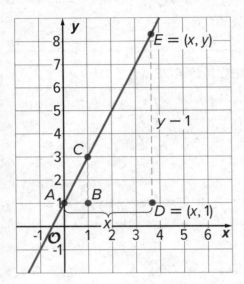

The slope for triangle ABC is $\frac{2}{1}$ since the vertical side has length 2 and the horizontal side has length 1.

For triangle ADE, the horizontal side has length x. The vertical side has length $y - 1$ because the distance from (x, y) to the x-axis is y but the vertical side of the triangle stops 1 unit short of the x-axis.

So the slope we find for triangle ADE is $\frac{y-1}{x}$.

The slopes for the two slope triangles are equal, meaning: $\frac{2}{1} = \frac{y-1}{x}$.

Since $y - 1$ is twice x, another way to write this equation is $y - 1 = 2x$.

This equation is true for any point (x, y) on the line!

NAME _____ DATE _____ PERIOD _____

Practice
Writing Equations for Lines

1. For each pair of points, find the slope of the line that passes through both points. If you get stuck, try plotting the points on graph paper and drawing the line through them with a ruler.

   a. (1, 1) and (7, 5)

   b. (1, 1) and (5, 7)

   c. (2, 5) and (-1, 2)

   d. (2, 5) and (-7, -4)

2. Line ℓ is shown in the coordinate plane.

   a. What are the coordinates of points B and D?

   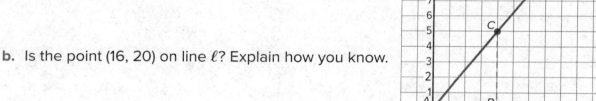

   b. Is the point (16, 20) on line ℓ? Explain how you know.

   c. Is the point (20, 24) on line ℓ? Explain how you know.

   d. Is the point (80, 100) on line ℓ? Explain how you know.

   e. Write a rule that would allow you to test whether (x, y) is on line ℓ.

3. Consider the graphed line.

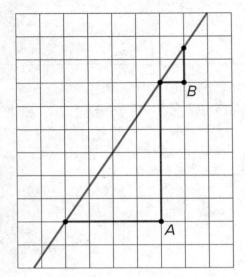

Mai uses Triangle A and says the slope of this line is $\frac{6}{4}$.

Elena uses Triangle B and says no, the slope of this line is 1.5.

Do you agree with either of them? Explain.

4. A rectangle has length 6 and height 4.

Which of these would tell you that quadrilateral *ABCD* is definitely *not* similar to this rectangle? Select **all** that apply. (Lesson 2-7)

(A.) $AB = BC$

(B.) $m\angle ABC = 105°$

(C.) $AB = 8$

(D.) $BC = 8$

(E.) $BC = 2 \cdot AB$

(F.) $2 \cdot AB = 3 \cdot BC$

Lesson 2-12

Using Equations for Lines

NAME _____ DATE _____ PERIOD _____

Learning Goal Let's write equations for lines.

Warm Up
12.1 Missing center

A dilation with scale factor 2 sends *A* to *B*. Where is the center of the dilation?

B

A

Activity
12.2 Writing Relationships from Two Points

Here is a line.

1. Using what you know about similar triangles, find an equation for the line in the diagram.

2. What is the slope of this line? Does it appear in your equation?

3. Is (9, 11) also on the line? How do you know?

4. Is (100, 193) also on the line?

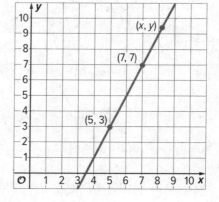

There are many different ways to write down an equation for a line like the one in the problem. Does $\dfrac{y-3}{x-6} = 2$ represent the line? What about $\dfrac{y-6}{x-4} = 5$?

What about $\dfrac{y+5}{x-1} = 2$? Explain your reasoning.

Activity

12.3 Dilations and Slope Triangles

Here is triangle *ABC*.

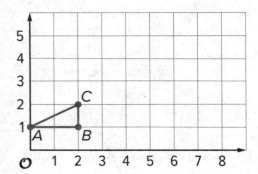

1. Draw the dilation of triangle *ABC* with center (0, 1) and scale factor 2.

2. Draw the dilation of triangle *ABC* with center (0, 1) and scale factor 2.5.

3. Where is *C* mapped by the dilation with center (0, 1) and scale factor *s*?

4. For which scale factor does the dilation with center (0, 1) send *C* to (9, 5.5)? Explain how you know.

NAME _____ DATE _____ PERIOD _____

Summary
Using Equations for Lines

We can use what we know about slope to decide if a point lies on a line. Here is a line with a few points labeled.

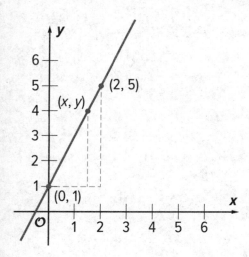

The slope triangle with vertices (0, 1) and (2, 5) gives a slope of $\frac{5-1}{2-0} = 2$.

The slope triangle with vertices (0, 1) and (x, y) gives a slope of $\frac{y-1}{x}$.

Since these slopes are the same, $\frac{y-1}{x} = 2$ is an equation for the line.

So, if we want to check whether or not the point (11, 23) lies on this line, we can check that $\frac{23-1}{11} = 2$.

Since (11, 23) is a solution to the equation, it is on the line!

Practice
Using Equations for Lines

1. Select **all** the points that are on the line through (0, 5) and (2, 8).

   (A.) (4, 11)

   (B.) (5, 10)

   (C.) (6, 14)

   (D.) (30, 50)

   (E.) (40, 60)

2. All three points displayed are on the line. Find an equation relating x and y.

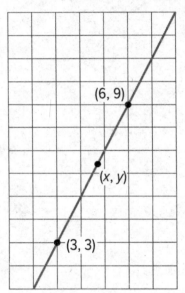

NAME _____ DATE _____ PERIOD _____

3. Here is triangle *ABC*.

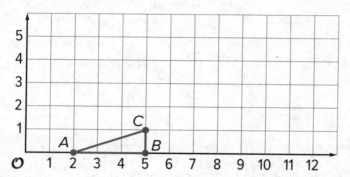

a. Draw the dilation of triangle *ABC* with center (2, 0) and scale factor 2.

b. Draw the dilation of triangle *ABC* with center (2, 0) and scale factor 3.

c. Draw the dilation of triangle *ABC* with center (2, 0) and scale factor $\frac{1}{2}$.

d. What are the coordinates of the image of point *C* when triangle *ABC* is dilated with center (2, 0) and scale factor *s*?

e. Write an equation for the line containing all possible images of point *C*.

4. Here are some line segments. (Lesson 2-4)

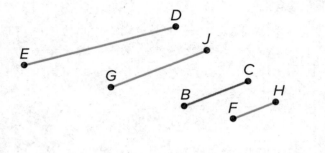

a. Which segment is a dilation of $\overline{BC}$ using A as the center of dilation and a scale factor of $\frac{2}{3}$?

b. Which segment is a dilation of $\overline{BC}$ using A as the center of dilation and a scale factor of $\frac{3}{2}$?

c. Which segment is not a dilation of $\overline{BC}$, and how do you know?

Lesson 2-13

The Shadow Knows

NAME _____ DATE _____ PERIOD _____

Learning Goal Let's use shadows to find the heights of an object.

Warm Up

13.1 Notice and Wonder: Long Shadows and Short Shadows

What do you notice? What do you wonder?

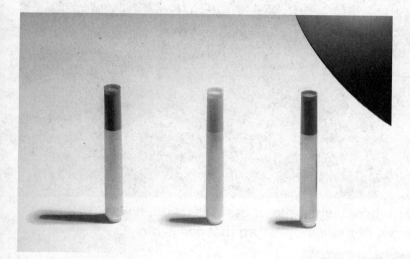

Here are some measurements that were taken when the photo was taken. It was impossible to directly measure the height of the lamppost, so that cell is blank.

	Height (inches)	Shadow Length (inches)
Younger Boy	43	29
Man	72	48
Older Boy	51	34
Lamppost		114

1. What relationships do you notice between an object's height and the length of its shadow?

2. Make a conjecture about the height of the lamppost and explain your thinking.

NAME _____ DATE _____ PERIOD _____

Activity
13.3 Justifying the Relationship

Explain *why* the relationship between the height of these objects and the length of their shadows is approximately proportional.

Activity

13.4 The Height of a Tall Object

1. Head outside. Make sure that it is a sunny day and you take a measuring device (like a tape measure or meter stick) as well as a pencil and some paper.

2. Choose an object whose height is too large to measure directly. Your teacher may assign you an object.

3. Use what you have learned to figure out the height of the object! Explain or show your reasoning.

Learning Targets

Lesson	Learning Target(s)
2-1 Projecting and Scaling	• I can decide if one rectangle is a dilation of another rectangle. • I know how to use a center and a scale factor to describe a dilation.
2-2 Circular Grid	• I can apply dilations to figures on a circular grid when the center of dilation is the center of the grid.
2-3 Dilations with No Grid	• I can apply a dilation to a polygon using a ruler.
2-4 Dilations on a Square Grid	• I can apply dilations to figures on a square grid. • If I know the angle measures and side lengths of a polygon, I know the angle measures and side lengths of the polygon if I apply a dilation with a certain scale factor.

(continued on the next page)

(continued from the previous page)

Lesson	Learning Target(s)
2-5 More Dilations	• I can apply dilations to polygons on a rectangular grid if I know the coordinates of the vertices and of the center of dilation.
2-6 Similarity	• I can apply a sequence of transformations to one figure to get a similar figure. • I can use a sequence of transformations to explain why two figures are similar.
2-7 Similar Polygons	• I can use angle measures and side lengths to conclude that two polygons are not similar. • I know the relationship between angle measures and side lengths in similar polygons.
2-8 Similar Triangles	• I know how to decide if two triangles are similar just by looking at their angle measures.

Lesson	Learning Target(s)
2-9 Side Length Quotients in Similar Triangles	• I can decide if two triangles are similar by looking at quotients of lengths of corresponding sides. • I can find missing side lengths in a pair of similar triangles using quotients of side lengths.
2-10 Meet Slope	• I can draw a line on a grid with a given slope. • I can find the slope of a line on a grid.
2-11 Writing Equations for Lines	• I can decide whether a point is on a line by finding quotients of horizontal and vertical distances.
2-12 Using Equations for Lines	• I can find an equation for a line and use that to decide which points are on that line.

(continued on the next page)

(continued from the previous page)

Lesson	Learning Target(s)
2-13 The Shadow Knows	• I can model a real-world context with similar triangles to find the height of an unknown object.

Notes:

_____

_____

_____

_____

_____

_____

_____

_____

_____

_____

_____

_____

_____

_____

_____

_____

_____

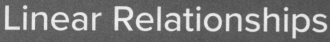

Unit 3

Linear Relationships

The steepness of a mountain or hill can be described as its *slope*.
You'll learn more about the slope of linear relationships in this unit.

Topics
- Proportional Relationships
- Representing Linear Relationships
- Finding Slopes
- Linear Equations
- Let's Put It to Work

Unit 3
Linear Relationships

Lesson 3-1

Understanding Proportional Relationships

NAME _____ DATE _____ PERIOD _____

Learning Goal Let's study some graphs.

Warm Up
1.1 Notice and Wonder: Two Graphs

What do you notice? What do you wonder?

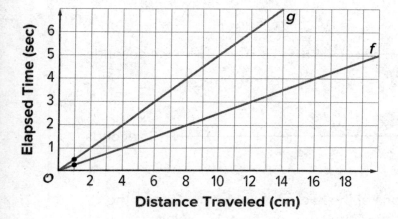

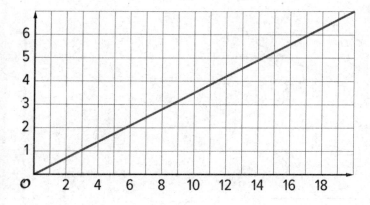

A ladybug and ant move at constant speeds. The diagrams with tick marks show their positions at different times. Each tick mark represents 1 centimeter.

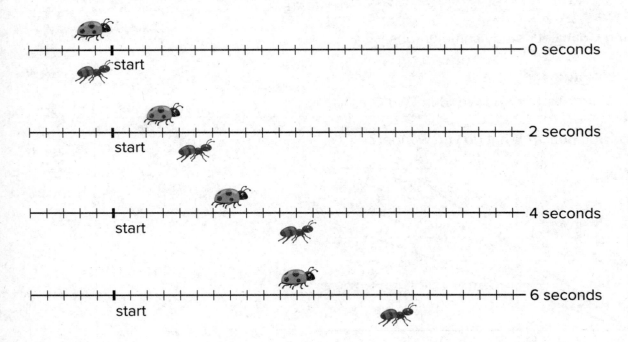

1. Lines *u* and *v* also show the positions of the two bugs. Which line shows the ladybug's movement? Which line shows the ant's movement? Explain your reasoning.

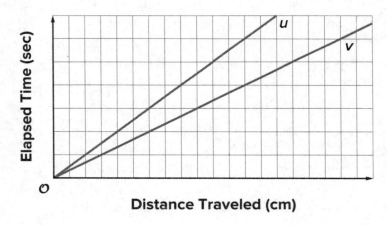

2. How long does it take the ladybug to travel 12 cm? The ant?

NAME _____ DATE _____ PERIOD _____

3. Scale the vertical and horizontal axes by labeling each grid line with a number. You will need to use the time and distance information shown in the tick-mark diagrams.

4. Mark and label the point on line u and the point on line v that represent the time and position of each bug after traveling 1 cm.

Are you ready for more?

1. How fast is each bug traveling?

2. Will there ever be a time when the ant is twice as far away from the start as the ladybug? Explain or show your reasoning.

Activity

1.3 Moving Twice as Fast

Refer to the tick-mark diagrams and graph in the earlier activity when needed.

1. Imagine a bug that is moving twice as fast as the ladybug. On each tick-mark diagram, mark the position of this bug.

2. Plot this bug's positions on the coordinate axes with lines u and v, and connect them with a line.

3. Write an equation for each of the three lines.

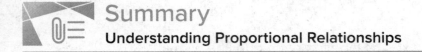

Summary
Understanding Proportional Relationships

Graphing is a way to help us make sense of relationships. But the graph of a line on a coordinate axis without scale or labels isn't very helpful.

For example, let's say we know that on longer bike rides Kiran can ride 4 miles every 16 minutes and Mai can ride 4 miles every 12 minutes.
Here are the graphs of these relationships.

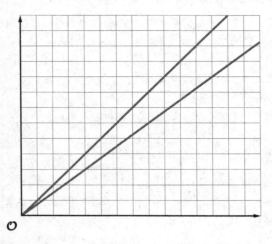

Without labels we can't even tell which line is Kiran and which is Mai!
Without labels and a scale on the axes, we can't use these graphs to answer questions like:

1. Which graph goes with which rider?

2. Who rides faster?

3. If Kiran and Mai start a bike trip at the same time, how far are they after 24 minutes?

4. How long will it take each of them to reach the end of the 12-mile bike path?

NAME _____ DATE _____ PERIOD _____

Here are the same graphs, but now with labels and scale.

Revisiting the questions from earlier:

1. Which graph goes with each rider?
   If Kiran rides 4 miles in 16 minutes, then the point (4, 16) is on his graph. If he rides for 1 mile, it will take 4 minutes. 10 miles will take 40 minutes. So, the upper graph represents Kiran's ride. Mai's points for the same distances are (1, 3), (4, 12), and (10, 30), so hers is the lower graph. (A letter next to each line would help us remember which is which!)

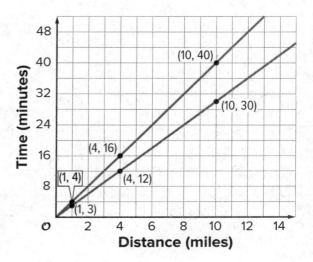

2. Who rides faster?
   Mai rides faster because she can ride the same distance as Kiran in a shorter time.

3. If Kiran and Mai start a bike trip at the same time, how far are they after 20 minutes?
   The points on the graphs at height 20 are 5 miles for Kiran and a little less than 7 miles for Mai.

4. How long will it take each of them to reach the end of the 12-mile bike path?
   The points on the graphs at a horizontal distance of 12 are 36 minutes for Mai and 48 minutes for Kiran. (Kiran's time after 12 miles is almost off the grid!)

Glossary

constant of proportionality

---

1. Priya jogs at a constant speed. The relationship between her distance and time is shown on the graph. Diego bikes at a constant speed twice as fast as Priya. Sketch a graph showing the relationship between Diego's distance and time.

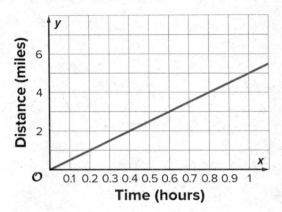

2. A you-pick blueberry farm offers 6 lbs of blueberries for $16.50.

   Sketch a graph of the relationship between cost and pounds of blueberries.

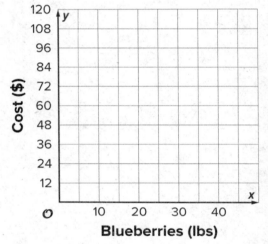

3. A line contains the points (-4, 1) and (4, 6). Decide whether or not each of these points is also on the line: (Lesson 2-12)

   a. (0, 3.5)   b. (12, 11)

   c. (80, 50)   d. (-1, 2.875)

4. The points (2, -4), (x, y), A, and B all lie on the line. Find an equation relating x and y. (Lesson 2-11)

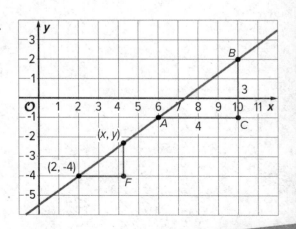

Lesson 3-2

Graphs of Proportional Relationships

NAME _____ DATE _____ PERIOD _____

Learning Goal Let's think about scale.

Warm Up
2.1 An Unknown Situation

Here is a graph that could represent a variety of different situations.

1. Write an equation for the graph.

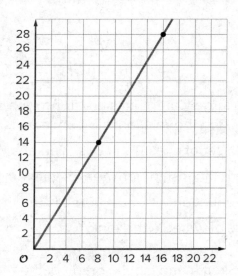

2. Sketch a new graph of this relationship.

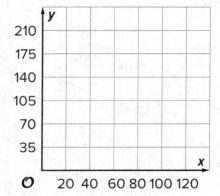

Activity

2.2 Card Sort: Proportional Relationships

Your teacher will give you 12 graphs of proportional relationships.

1. Sort the graphs into groups based on what proportional relationship they represent.

2. Write an equation for each *different* proportional relationship you find.

Activity

2.3 Different Scales

Two large water tanks are filling with water. Tank A is not filled at a constant rate, and the relationship between its volume of water and time is graphed on each set of axes. Tank B is filled at a constant rate of $\frac{1}{2}$ liters per minute. The relationship between its volume of water and time can be described by the equation $v = \frac{1}{2}t$, where t is the time in minutes and v is the total volume in liters of water in the tank.

1. Sketch and label a graph of the relationship between the volume of water v and time t for Tank B on each of the axes.

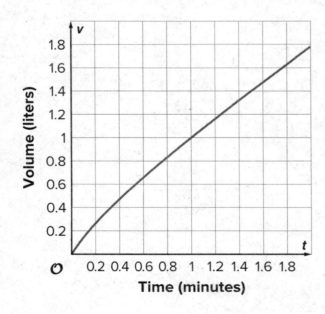

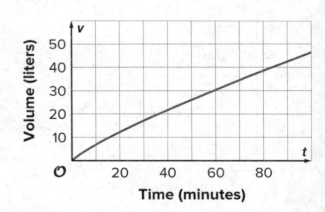

NAME _____ DATE _____ PERIOD _____

2. Answer the following questions and say which graph you used to find your answer.

   a. After 30 seconds, which tank has the most water?

   b. At approximately what times do both tanks have the same amount of water?

   c. At approximately what times do both tanks contain 1 liter of water? 20 liters?

Are you ready for more?

A giant tortoise travels at 0.17 miles per hour and an arctic hare travels at 37 miles per hour.

1. Draw separate graphs that show the relationship between time elapsed, in hours, and distance traveled, in miles, for both the tortoise and the hare.

2. Would it be helpful to try to put both graphs on the same pair of axes? Why or why not?

3. The tortoise and the hare start out together and after half an hour the hare stops to take a rest. How long does it take the tortoise to catch up?

The scales we choose when graphing a relationship often depend on what information we want to know.

For example, say two water tanks are filled at different constant rates. The relationship between time in minutes t and volume in liters v of Tank A is given by $v = 2.2t$. For Tank B the relationship is $v = 2.75t$.

These equations tell us that Tank A is being filled at a constant rate of 2.2 liters per minute and Tank B is being filled at a constant rate of 2.75 liters per minute.

If we want to use graphs to see at what times the two tanks will have 110 liters of water, then using an axis scale from 0 to 10, as shown here, isn't very helpful.

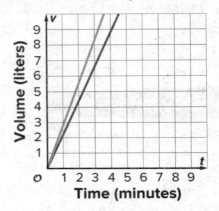

If we use a vertical scale that goes to 150 liters, a bit beyond the 110 we are looking for, and a horizontal scale that goes to 100 minutes, we get a much more useful set of axes for answering our question.

Now we can see that the two tanks will reach 110 liters 10 minutes apart—Tank B after 40 minutes of filling and Tank A after 50 minutes of filling.

It is important to note that both of these graphs are correct, but one uses a range of values that helps answer the question.

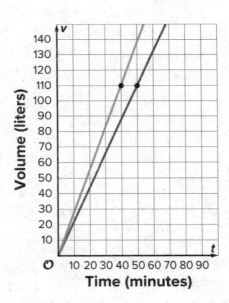

In order to always pick a helpful scale, we should consider the situation and the questions asked about it.

NAME _____ DATE _____ PERIOD _____

Practice
Graphs of Proportional Relationships

1. The tortoise and the hare are having a race. After the hare runs 16 miles the tortoise has only run 4 miles.

   The relationship between the distance *x* the tortoise "runs" in miles for every *y* miles the hare runs is $y = 4x$. Graph this relationship.

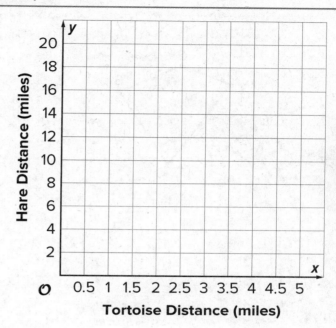

2. The table shows a proportional relationship between the weight on a spring scale and the distance the spring has stretched.

Distance (cm)	Weight (newtons)
20	28
55	
	140
1	

   a. Complete the table.

   b. Describe the scales you could use on the *x* and *y* axes of a coordinate grid that would show all the distances and weights in the table.

3. Find a sequence of rotations, reflections, translations, and dilations showing that one figure is similar to the other. Be specific: give the amount and direction of a translation, a line of reflection, the center and angle of a rotation, and the center and scale factor of a dilation. (Lesson 2-6)

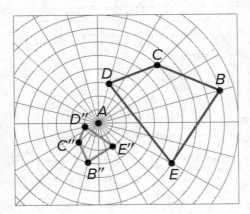

4. Andre said, "I found two figures that are congruent, so they can't be similar." Diego said, "No, they are similar! The scale factor is 1." Do you agree with either of them? Use the definition of similarity to explain your answer. (Lesson 2-6)

Lesson 3-3

Representing Proportional Relationships

NAME _____ DATE _____ PERIOD _____

Learning Goal Let's graph proportional relationships.

Warm Up
3.1 Number Talk: Multiplication

Find the value of each product mentally.

1. $15 \cdot 2$ 2. $15 \cdot 0.5$ 3. $15 \cdot 0.25$ 4. $15 \cdot (2.25)$

Activity
3.2 Representations of Proportional Relationships

1. Here are two ways to represent a situation.

Description: Jada and Noah counted the number of steps they took to walk a set distance. To walk the same distance, Jada took 8 steps while Noah took 10 steps. Then they found that when Noah took 15 steps, Jada took 12 steps.

Equation: Let x represent the number of steps Jada takes and let y represent the number of steps Noah takes.

$$y = \frac{5}{4}x$$

a. Create a table that represents this situation with at least 3 pairs of values.

b. Graph this relationship and label the axes.

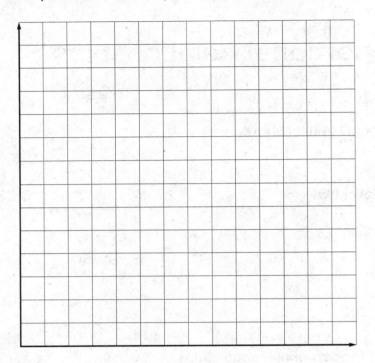

c. How can you see or calculate the constant of proportionality in each representation? What does it mean?

d. Explain how you can tell that the equation, description, graph, and table all represent the same situation.

NAME _____ DATE _____ PERIOD _____

2. Here are two ways to represent a situation.

Description: The Origami Club is doing a car wash fundraiser to raise money for a trip. They charge the same price for every car. After 11 cars, they raised a total of $93.50. After 23 cars, they raised a total of $195.50.

Table:

Number of Cars	Amount Raised in Dollars
11	93.50
23	195.50

a. Write an equation that represents this situation. (Use c to represent number of cars and use m to represent amount raised in dollars.)

b. Create a graph that represents this situation.

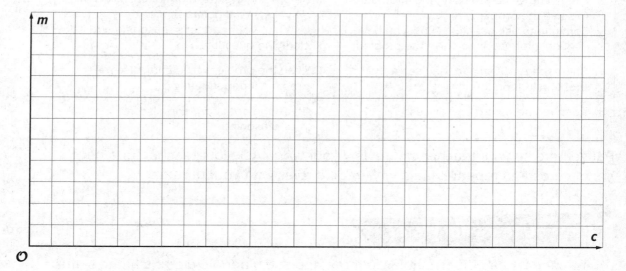

c. How can you see or calculate the constant of proportionality in each representation? What does it mean?

d. Explain how you can tell that the equation, description, graph, and table all represent the same situation.

Activity

3.3 Info Gap: Proportional Relationships

Your teacher will give you either a *problem card* or a *data card*. Do not show or read your card to your partner.

If your teacher gives you the *problem card...*	If your teacher gives you the *data card...*
1. Silently read your card and think about what information you need to be able to answer the question.	1. Silently read your card.
2. Ask your partner for the specific information that you need.	2. Ask your partner *"What specific information do you need?"* and wait for them to *ask* for information. If your partner asks for information that is not on the card, do not do the calculations for them. Tell them you don't have that information.
3. Explain how you are using the information to solve the problem. Continue to ask questions until you have enough information to solve the problem.	3. Before sharing the information, ask *"Why do you need that information?"* Listen to your partner's reasoning and ask clarifying questions.
4. Share the *problem card* and solve the problem independently.	4. Read the *problem card* and solve the problem independently.
5. Read the *data card* and discuss your reasoning.	5. Share the *data card* and discuss your reasoning.

Pause here so your teacher can review your work. Ask your teacher for a new set of cards and repeat the activity, trading roles with your partner.

Are you ready for more?

Ten people can dig five holes in three hours. If n people digging at the same rate dig m holes in d hours:

1. Is n proportional to m when $d = 3$?

2. Is n proportional to d when $m = 5$?

3. Is m proportional to d when $n = 10$?

NAME _____ DATE _____ PERIOD _____

Summary
Representing Proportional Relationships

Proportional relationships can be represented in multiple ways. Which representation we choose depends on the purpose. And when we create representations we can choose helpful values by paying attention to the context.

For example, a stew recipe calls for 3 carrots for every 2 potatoes. One way to represent this is using an equation. If there are p potatoes and c carrots, then $c = \frac{3}{2}p$.

Suppose we want to make a large batch of this recipe for a family gathering, using 150 potatoes. To find the number of carrots we could just use the equation: $\frac{3}{2} \cdot 150 = 225$ carrots.

Now suppose the recipe is used in a restaurant that makes the stew in large batches of different sizes depending on how busy a day it is, using up to 300 potatoes at a time.

- Then we might make a graph to show how many carrots are needed for different amounts of potatoes.

- We set up a pair of coordinate axes with a scale from 0 to 300 along the horizontal axis and 0 to 450 on the vertical axis, because $450 = \frac{3}{2} \cdot 300$.

- Then we can read how many carrots are needed for any number of potatoes up to 300.

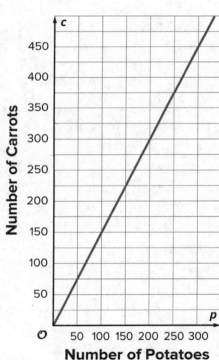

Or if the recipe is used in a food factory that produces very large quantities and the potatoes come in bags of 150, we might just make a table of values showing the number of carrots needed for different multiples of 150.

Number of Potatoes	Number of Carrots
150	225
300	450
450	675
600	900

No matter the representation or the scale used, the constant of proportionality, $\frac{3}{2}$, is evident in each. In the equation it is the number we multiply p by; in the graph, it is the slope; and in the table, it is the number we multiply by values in the left column to get numbers in the right column. We can think of the constant of proportionality as a **rate of change** of c with respect to p. In this case, the rate of change is $\frac{3}{2}$ carrots per potato.

Glossary

rate of change

NAME _____ DATE _____ PERIOD _____

Practice
Representing Proportional Relationships

1. Here is a graph of the proportional relationship between calories and grams of fish.

   a. Write an equation that reflects this relationship using x to represent the amount of fish in grams and y to represent the number of calories.

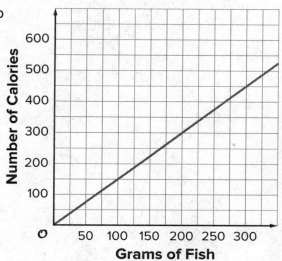

   b. Use your equation to complete the table:

Grams of Fish	Number of Calories
1,000	
	2,001
1	

2. Students are selling raffle tickets for a school fundraiser. They collect $24 for every 10 raffle tickets they sell.

   a. Suppose M is the amount of money the students collect for selling R raffle tickets. Write an equation that reflects the relationship between M and R.

b. Label and scale the axes and graph this situation with M on the vertical axis and R on the horizontal axis. Make sure the scale is large enough to see how much they would raise if they sell 1,000 tickets.

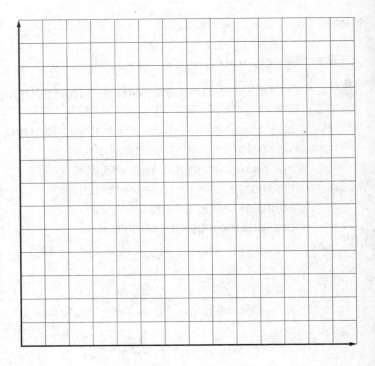

3. Describe how you can tell whether a line's slope is greater than 1, equal to 1, or less than 1. **(Lesson 2-10)**

4. A line is represented by the equation $\frac{y}{x-2} = \frac{3}{11}$. What are the coordinates of some points that lie on the line? Graph the line on graph paper. **(Lesson 2-12)**

Lesson 3-4

Comparing Proportional Relationships

NAME _____ DATE _____ PERIOD _____

Learning Goal Let's compare proportional relationships.

 ## Warm Up
4.1 What's the Relationship?

The equation $y = 4.2x$ could represent a variety of different situations.

1. Write a description of a situation represented by this equation. Decide what quantities x and y represent in your situation.

2. Make a table and a graph that represent the situation.

Activity

4.2 Comparing Two Different Representations

1. Elena babysits her neighbor's children. Her earnings are given by the equation $y = 8.40x$, where x represents the number of hours she worked and y represents the amount of money she earned. Jada earns $7 per hour mowing her neighbors' lawns.

   a. Who makes more money after working 12 hours? How much more do they make? Explain your reasoning by creating a graph or a table.

   b. What is the rate of change for each situation and what does it mean?

   c. Using your graph or table, determine how long it would take each person to earn $150.

2. Clare and Han have summer jobs stuffing envelopes for two different companies. Clare's earnings can be seen in the table. Han earns $15 for every 300 envelopes he finishes.

Number of Envelopes	Money in Dollars
400	40
900	90

   a. By creating a graph, show how much money each person makes after stuffing 1,500 envelopes.

NAME _____ DATE _____ PERIOD _____

b. What is the rate of change for each situation and what does it mean?

c. Using your graph, determine how much more money one person makes relative to the other after stuffing 1,500 envelopes. Explain or show your reasoning.

3. Tyler plans to start a lemonade stand and is trying to perfect his recipe for lemonade. He wants to make sure the recipe doesn't use too much lemonade mix (lemon juice and sugar) but still tastes good. Lemonade Recipe 1 is given by the equation $y = 4x$ where x represents the amount of lemonade mix in cups and y represents the amount of water in cups. Lemonade Recipe 2 is given in the table.

Lemonade Mix (cups)	Water (cups)
10	50
13	65
21	105

a. If Tyler had 16 cups of lemonade mix, how many cups of water would he need for each recipe? Explain your reasoning by creating a graph or a table.

b. What is the rate of change for each situation and what does it mean?

c. Tyler has a 5-gallon jug (which holds 80 cups) to use for his lemonade stand and 16 cups of lemonade mix. Which lemonade recipe should he use? Explain or show your reasoning.

Han and Clare are still stuffing envelopes. Han can stuff 20 envelopes in a minute, and Clare can stuff 10 envelopes in a minute. They start working together on a pile of 1,000 envelopes.

1. How long does it take them to finish the pile?

2. Who earns more money?

Summary
Comparing Proportional Relationships

When two proportional relationships are represented in different ways, we compare them by finding a common piece of information.

For example, Clare's earnings are represented by the equation $y = 14.5x$, where y is her earnings in dollars for working x hours.

The table shows some information about Jada's pay.

Time Worked (hours)	Earnings (dollars)
7	92.75
4.5	59.63
37	490.25

Who is paid at a higher rate per hour? How much more does that person have after 20 hours?

In Clare's equation we see that the rate of change (how many dollars she earns every hour) is 14.50.

We can calculate Jada's rate of change by dividing a value in the earnings column by the value in the same row in the time worked column. Using the last row, the rate of change for Jada is 13.25, since $490.25 \div 37 = 13.25$. An equation representing Jada's earnings is $y = 13.25x$. This means she earns $13.25 per hour.

So, Clare is paid at a higher rate than Jada. Clare earns $1.25 more per hour than Jada. After 20 hours of work, she earns $25 more than Jada because $20 \cdot (1.25) = 25$.

NAME _____ DATE _____ PERIOD _____

Practice
Comparing Proportional Relationships

1. A contractor must haul a large amount of dirt to a work site. She collected information from two hauling companies. EZ Excavation gives its prices in a table. Happy Hauling Service gives its prices in a graph.

Dirt (cubic yards)	Cost (dollars)
8	196
20	490
26	637

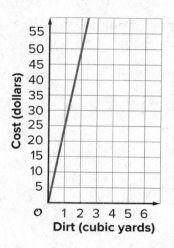

a. How much would each hauling company charge to haul 40 cubic yards of dirt? Explain or show your reasoning.

b. Calculate the rate of change for each relationship. What do they mean for each company?

c. If the contractor has 40 cubic yards of dirt to haul and a budget of $1,000, which hauling company should she hire? Explain or show your reasoning.

2. Andre and Priya are tracking the number of steps they walk. Andre records that he can walk 6,000 steps in 50 minutes. Priya writes the equation $y = 118x$, where y is the number of steps and x is the number of minutes she walks, to describe her step rate. This week, Andre and Priya each walk for a total of 5 hours. Who walks more steps? How many more?

3. Find the coordinates of point D in each diagram: (Lesson 2-11)

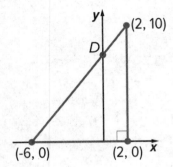

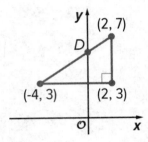

4. Select **all** the pairs of points so that the line between those points has slope $\frac{2}{3}$. (Lesson 2-11)

A. (0, 0) and (2, 3)

B. (0, 0) and (3, 2)

C. (1, 5) and (4, 7)

D. (-2, -2) and (4, 2)

E. (20, 30) and (-20, -30)

Lesson 3-5

Introduction to Linear Relationships

NAME _____ DATE _____ PERIOD _____

Learning Goal Let's explore some relationships between two variables.

Warm Up
5.1 Number Talk: Fraction Division

Find the value of $2\frac{5}{8} \div \frac{1}{2}$.

Activity
5.2 Stacking Cups

We have two stacks of styrofoam cups.

- One stack has 6 cups, and its height is 15 cm.

- The other stack has 12 cups, and its height is 23 cm.

How many cups are needed for a stack with a height of 50 cm?

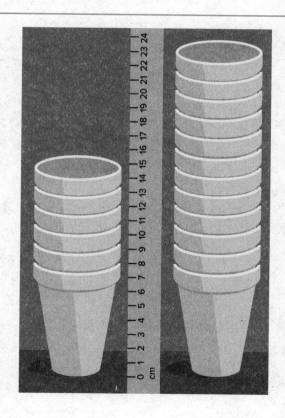

5.3 Connecting Slope to Rate of Change

1. If you didn't create your own graph of the situation before, do so now.

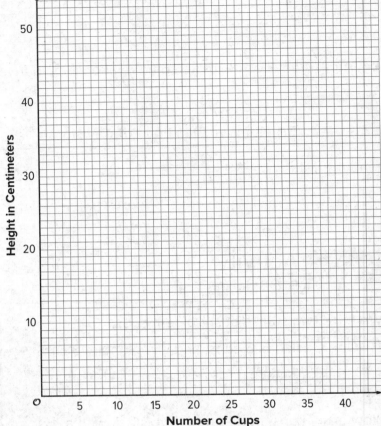

2. What are some ways you can tell that the number of cups is not proportional to the height of the stack?

3. What is the **slope** of the line in your graph? What does the slope mean in this situation?

4. At what point does your line intersect the vertical axis? What do the coordinates of this point tell you about the cups?

5. How much height does each cup, after the first, add to the stack?

NAME _____ DATE _____ PERIOD _____

Summary
Introduction to Linear Relationships

Andre starts babysitting and charges $10 for traveling to and from the job, and $15 per hour. For every additional hour he works, he charges another $15. If we graph Andre's earnings based on how long he works, we have a line that starts at $10 on the vertical axis and then increases by $15 each hour.

A **linear relationship** is any relationship between two quantities where one quantity has a constant **rate of change** with respect to the other.

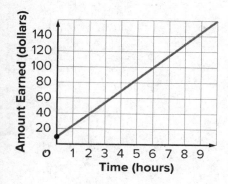

We can figure out the rate of change using the graph. Because the rate of change is constant, we can take any two points on the graph and divide the amount of vertical change by the amount of horizontal change.

For example, take the points (2, 40) and (6, 100). They mean that Andre earns $40 for working 2 hours and $100 for working 6 hours. The rate of change is $\frac{100-40}{6-2} = 15$ dollars per hour. Andre's earnings go up $15 for each hour of babysitting.

Notice that this is the same way we calculate the **slope** of the line. That's why the graph is a line, and why we call this a linear relationship. The rate of change of a linear relationship is the same as the slope of its graph.

With proportional relationships we are used to graphs that contain the point (0, 0). But proportional relationships are just one type of linear relationship. In the following lessons, we will continue to explore the other type of linear relationship where the quantities are not both 0 at the same time.

Glossary

linear relationship

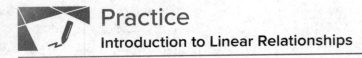

Practice
Introduction to Linear Relationships

1. A restaurant offers delivery for their pizzas. The total cost is a delivery fee added to the price of the pizzas. One customer pays $25 to have 2 pizzas delivered. Another customer pays $58 for 5 pizzas. How many pizzas are delivered to a customer who pays $80?

2. To paint a house, a painting company charges a flat rate of $500 for supplies, plus $50 for each hour of labor.

   a. How much would the painting company charge to paint a house that needs 20 hours of labor? A house that needs 50 hours?

NAME _____ DATE _____ PERIOD _____

b. Draw a line representing the relationship between x, the number of hours it takes the painting company to finish the house, and y, the total cost of painting the house. Label the two points from the earlier question on your graph.

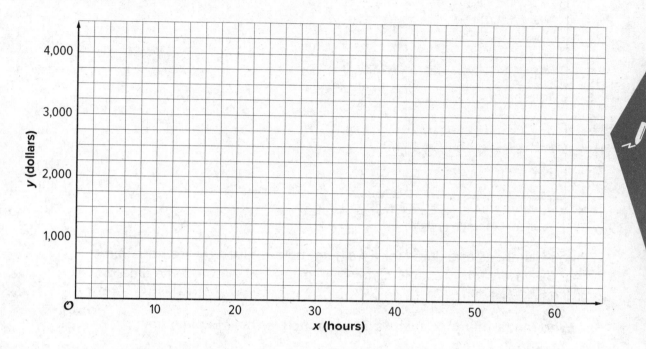

c. Find the slope of the line. What is the meaning of the slope in this context?

3. Tyler and Elena are on the cross-country team. Tyler's distances and times for a training run are shown on the graph.

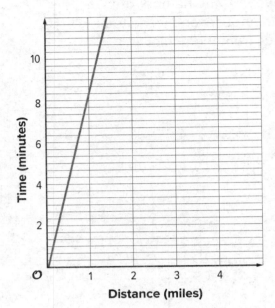

Elena's distances and times for a training run are given by the equation $y = 8.5x$, where x represents distance in miles and y represents time in minutes. (Lesson 3-4)

a. Who ran farther in 10 minutes? How much farther? Explain how you know.

b. Calculate each runner's pace in minutes per mile.

c. Who ran faster during the training run? Explain or show your reasoning.

4. Write an equation for the line that passes through (2, 5) and (6, 7).
   (Lesson 2-12)

Lesson 3-6

More Linear Relationships

NAME _____ DATE _____ PERIOD _____

Learning Goal Let's explore some more relationships between two variables.

Warm Up
6.1 Growing

Look for a growing pattern. Describe the pattern you see.

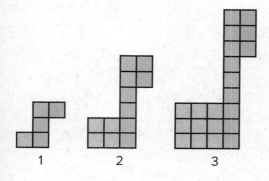

1. If your pattern continues growing in the same way, how many tiles of each color will be in the 4th and 5th diagram? The 10th diagram?

2. How many tiles of each color will be in the nth diagram? Be prepared to explain your reasoning.

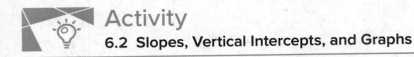

Activity

6.2 Slopes, Vertical Intercepts, and Graphs

Your teacher will give you 6 cards describing different situations and 6 cards with graphs.

1. Match each situation to a graph.

2. Pick one proportional relationship and one non-proportional relationship and answer the following questions about them.

   a. How can you find the slope from the graph? Explain or show your reasoning.

   b. Explain what the slope means in the situation.

   c. Find the point where the line crosses the vertical axis. What does that point tell you about the situation?

NAME _____ DATE _____ PERIOD _____

Activity

6.3 Summer Reading

Lin has a summer reading assignment. After reading the first 30 pages of the book, she plans to read 40 pages each day until she finishes. Lin makes the graph shown here to track how many total pages she'll read over the next few days.

After day 1, Lin reaches page 70, which matches the point (1, 70) she made on her graph. After day 4, Lin reaches page 190, which does not match the point (4, 160) she made on her graph. Lin is not sure what went wrong since she knows she followed her reading plan.

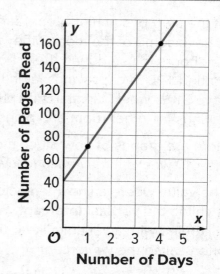

1. Sketch a line showing Lin's original plan on the axes.

2. What does the **vertical intercept** mean in this situation? How do the vertical intercepts of the two lines compare?

3. What does the slope mean in this situation? How do the slopes of the two lines compare?

Are you ready for more?

Jada's grandparents started a savings account for her in 2010. The table shows the amount in the account each year.

If this relationship is graphed with the year on the horizontal axis and the amount in dollars on the vertical axis, what is the vertical intercept? What does it mean in this context?

Year	Amount in Dollars
2010	600
2012	750
2014	900
2016	1,050

At the start of summer break, Jada and Lin decide to save some of the money they earn helping out their neighbors to use during the school year. Jada starts by putting $20 into a savings jar in her room and plans to save $10 a week. Lin starts by putting $10 into a savings jar in her room and plans to save $20 a week.

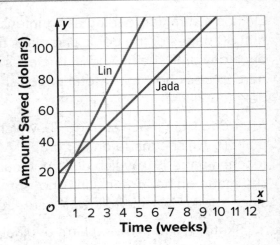

Here are graphs of how much money they will save after 10 weeks if they each follow their plans.

The value where a line intersects the vertical axis is called the **vertical intercept**. When the vertical axis is labeled with a variable like y, this value is also often called the *y-intercept*.

Jada's graph has a vertical intercept of $20 while Lin's graph has a vertical intercept of $10. These values reflect the amount of money they each started with.

- At 1 week they will have saved the same amount, $30.

- But after week 1, Lin is saving more money per week (so she has a larger rate of change), so she will end up saving more money over the summer if they each follow their plans.

Glossary

vertical intercept

NAME _____ DATE _____ PERIOD _____

Practice
More Linear Relationships

1. Explain what the slope and intercept mean in each situation.

   a. A graph represents the perimeter, y, in units, for an equilateral triangle with side length x units. The slope of the line is 3 and the y-intercept is 0.

   b. The amount of money, y, in a cash box after x tickets are purchased for carnival games. The slope of the line is $\frac{1}{4}$ and the y-intercept is 8.

   c. The number of chapters read, y, after x days. The slope of the line is $\frac{5}{4}$ and the y-intercept is 2.

   d. The graph shows the cost in dollars, y, of a muffin delivery and the number of muffins, x, ordered. The slope of the line is 2 and the y-intercept is 3.

2. Customers at the gym pay a membership fee to join and then a fee for each class they attend. Here is a graph that represents the situation.

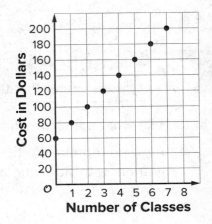

a. What does the slope of the line shown by the points mean in this situation?

b. What does the vertical intercept mean in this situation?

3. The graph shows the relationship between the number of cups of flour and the number of cups of sugar in Lin's favorite brownie recipe. The table shows the amounts of flour and sugar needed for Noah's favorite brownie recipe. (Lesson 3-4)

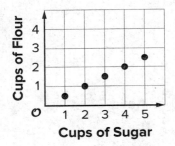

Cups of Sugar	Cups of Flour
$\frac{3}{2}$	1
3	2
$4\frac{1}{2}$	3

a. Noah and Lin buy a 12-cup bag of sugar and divide it evenly to make their recipes. If they each use all their sugar, how much flour do they each need?

b. Noah and Lin buy a 10-cup bag of flour and divide it evenly to make their recipes. If they each use all their flour, how much sugar do they each need?

Lesson 3-7

Representations of Linear Relationships

NAME _____ DATE _____ PERIOD _____

Learning Goal Let's write equations from real situations.

Warm Up
7.1 Estimation: Which Holds More?

Which glass will hold the most water? The least?

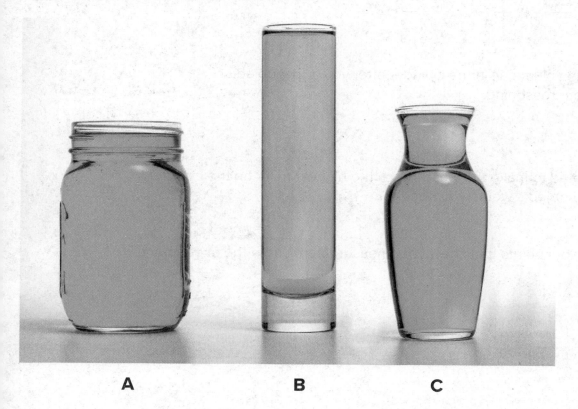

A	**B**	**C**

Activity

7.2 Rising Water Levels

1. Record data from your teacher's demonstration in the table. (You may not need all the rows.)

Number of Objects	Volume in ml

2. What is the volume, V, in the cylinder after you add x objects? Explain your reasoning.

3. If you wanted to make the water reach the highest mark on the cylinder, how many objects would you need?

4. Plot and label points that show your measurements from the experiment.

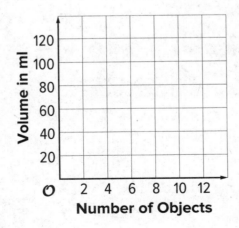

5. The points should fall on a line. Use a ruler to graph this line.

NAME _____ DATE _____ PERIOD _____

6. Compute the slope of the line. What does the slope mean in this situation?

7. What is the vertical intercept? What does the vertical intercept mean in this situation?

Are you ready for more?

A situation is represented by the equation $y = 5 + \frac{1}{2}x$.

1. Invent a story for this situation.

2. Graph the equation.

3. What do the $\frac{1}{2}$ and the 5 represent in your situation?

4. Where do you see the $\frac{1}{2}$ and 5 on the graph?

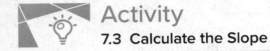
1. Refer to the following graphs.

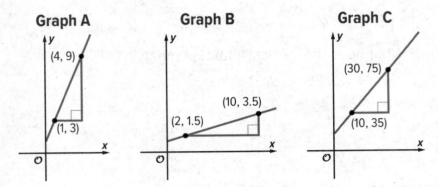

Graph A

(4, 9)

(1, 3)

Graph B

(10, 3.5)

(2, 1.5)

Graph C

(30, 75)

(10, 35)

For each graph, record:

Vertical Change	Horizontal Change	Slope

2. Describe a procedure for finding the slope between any two points on a line.

3. Write an expression for the slope of the line in the graph using the letters *u*, *v*, *s*, and *t*.

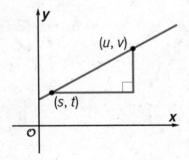

(u, v)

(s, t)

NAME _____ DATE _____ PERIOD _____

Summary
Representations of Linear Relationships

Let's say we have a glass cylinder filled with 50 ml of water and a bunch of marbles that are 3 ml in volume. If we drop marbles into the cylinder one at a time, we can watch the height of the water increase by the same amount, 3 ml, for each one added.

This constant rate of change means there is a linear relationship between the number of marbles and the height of the water.

• Add one marble, the water height goes up 3 ml.

• Add 2 marbles, the water height goes up 6 ml. Add x marbles, the water height goes up $3x$ ml.

Reasoning this way, we can calculate that the height, y, of the water for x marbles is $y = 3x + 50$. Any linear relationships can be expressed in the form $y = mx + b$ using just the rate of change, m, and the initial amount, b. The 3 represents the rate of change, or slope of the graph, and the 50 represents the initial amount, or vertical intercept of the graph. We'll learn about some more ways to think about this equation in future lessons.

Now what if we didn't have a description to use to figure out the slope and the vertical intercept? That's okay so long as we can find some points on the line!

For the line graphed here, two of the points on the line are (3, 3) and (9, 5) and we can use these points to draw in a slope triangle as shown.

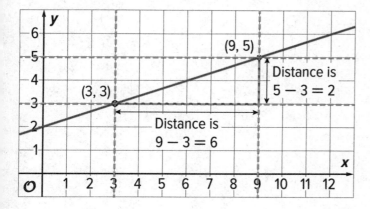

The slope of this line is the quotient of the length of the vertical side of the slope triangle and the length of the horizontal side of the slope triangle.

So the slope, m, is $\dfrac{\text{vertical change}}{\text{horizontal change}} = \dfrac{2}{6} = \dfrac{1}{3}$.

We can also see from the graph that the vertical intercept, b, is 2.

Putting these together, we can say that the equation for this line is $y = \dfrac{1}{3}x + 2$.

Practice
Representations of Linear Relationships

1. Create a graph that shows three linear relationships with different *y*-intercepts using the following slopes, and write an equation for each line.

Slopes

a. $\frac{1}{5}$

b. $\frac{3}{5}$

c. $\frac{6}{5}$

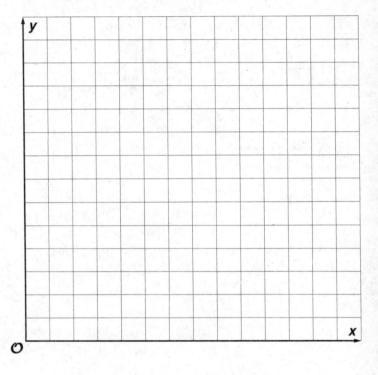

2. The graph shows the height in inches, *h*, of a bamboo plant *t* months after it has been planted.

a. Write an equation that describes the relationship between *h* and *t*.

b. After how many months will the bamboo plant be 66 inches tall? Explain or show your reasoning.

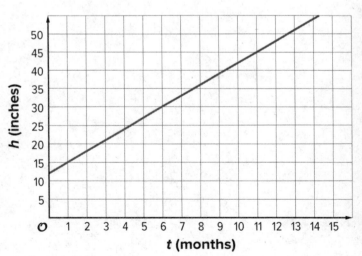

NAME _____ DATE _____ PERIOD _____

3. Here are recipes for two different banana cakes. Information for the first recipe is shown in the table. (Lesson 3-4)

Sugar (cups)	Flour (cups)
$\frac{1}{2}$	$\frac{3}{4}$
$2\frac{1}{2}$	$3\frac{3}{4}$
3	$4\frac{1}{2}$

The relationship between cups of flour y and cups of sugar x in the second recipe is $y = \frac{7}{4}x$.

a. If you used 4 cups of sugar, how much flour does each recipe need?

b. What is the constant of proportionality for each situation and what does it mean?

4. Show that the two figures are similar by identifying a sequence of translations, rotations, reflections, and dilations that takes the larger figure to the smaller one. (Lesson 2-6)

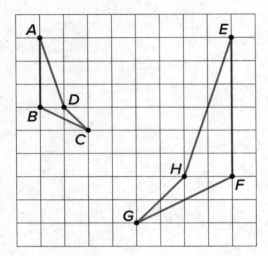

Lesson 3-8

Translating to $y = mx + b$

NAME _____ DATE _____ PERIOD _____

Learning Goal Let's see what happens to the equations of translated lines.

Warm Up
8.1 Lines that Are Translations

The diagram shows several lines. You can only see part of the lines, but they actually continue forever in both directions.

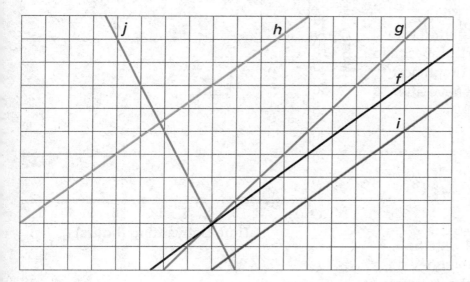

1. Which lines are images of line *f* under a translation?

2. For each line that is a translation of *f*, draw an arrow on the grid that shows the vertical translation distance.

1. Diego earns $10 per hour babysitting. Assume that he has no money saved before he starts babysitting and plans to save all of his earnings. Graph how much money, *y*, he has after *x* hours of babysitting.

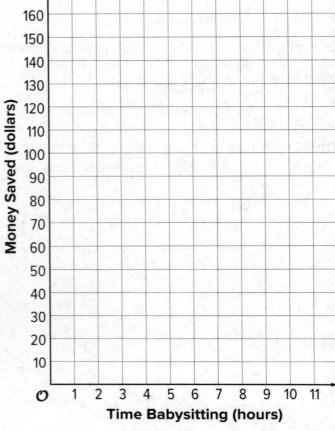

2. Now imagine that Diego started with $30 saved before he starts babysitting. On the same set of axes, graph how much money, *y*, he would have after *x* hours of babysitting.

3. Compare the second line with the first line. How much *more* money does Diego have after 1 hour of babysitting? 2 hours? 5 hours? *x* hours?

4. Write an equation for each line.

NAME _____ DATE _____ PERIOD _____

Activity

8.3 Translating a Line

This graph shows two lines. Line *a* goes through the origin (0, 0). Line *h* is the image of line *a* under a translation.

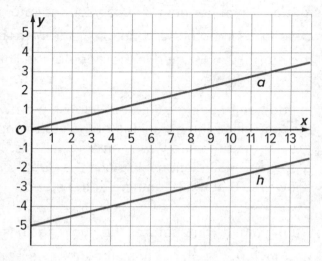

1. Select all of the equations whose graph is the line *h*.

   a. $y = \frac{1}{4}x - 5$

   b. $y = \frac{1}{4}x + 5$

   c. $\frac{1}{4}x - 5 = y$

   d. $y = -5 + \frac{1}{4}x$

   e. $-5 + \frac{1}{4}x = y$

   f. $y = 5 - \frac{1}{4}x$

2. Your teacher will give you 12 cards. There are 4 pairs of lines, A–D, showing the graph, *a*, of a proportional relationship and the image, *h*, of *a* under a translation. Match each line *h* with an equation and either a table or description. For the line with no matching equation, write one on the blank card.

Are you ready for more?

A student says that the graph of the equation $y = 3(x + 8)$ is the same as the graph of $y = 3x$, only translated upwards by 8 units. Do you agree? Why or why not?

During an early winter storm, the snow fell at a rate of $\frac{1}{2}$ inch per hour. We can see the rate of change, $\frac{1}{2}$, in both the equation that represents this storm, $y = \frac{1}{2}x$, and in the slope of the line representing this storm.

In addition to being a linear relationship between the time since the beginning of the storm and the depth of the snow, we can also call this a proportional relationship since the depth of snow was 0 at the beginning of the storm.

During a mid-winter storm, the snow again fell at a rate of $\frac{1}{2}$ inch per hour, but this time there was already 5 inches of snow on the ground. We can graph this storm on the same axes as the first storm by taking all the points on the graph of the first storm and translating them up 5 inches.

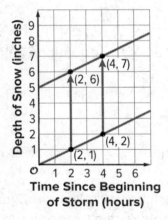

Two hours after each storm begins, 1 inch of new snow has fallen.

- For the first storm, this means there is now 1 inch of snow on the ground.

- For the second storm, this means there are now 6 inches of snow on the ground.

Unlike the first storm, the second is not a proportional relationship since the line representing the second storm has a vertical intercept of 5. The equation representing the storm, $y = \frac{1}{2}x + 5$, is of the form $y = mx + b$, where m is the rate of change, also the slope of the graph, and b is the initial amount, also the vertical intercept of the graph.

NAME _____ DATE _____ PERIOD _____

 ## Practice
Translating to $y = mx + b$

1. Select **all** the equations that have graphs with the same y-intercept.

   (A.) $y = 3x - 8$

   (B.) $y = 3x - 9$

   (C.) $y = 3x + 8$

   (D.) $y = 5x - 8$

   (E.) $y = 2x - 8$

   (F.) $y = \frac{1}{3}x - 8$

2. Create a graph showing the equations $y = \frac{1}{4}x$ and $y = \frac{1}{4}x - 5$. Explain how the graphs are the same and how they are different.

3. A cable company charges $70 per month for cable service to existing customers.

   a. Find a linear equation representing the relationship between x, the number of months of service, and y, the total amount paid in dollars by an existing customer.

   b. For new customers, there is an additional one-time $100 service fee. Repeat the previous problem for new customers.

   c. When the two equations are graphed in the coordinate plane, how are they related to each other geometrically?

NAME _____ DATE _____ PERIOD _____

4. A mountain road is 5 miles long and gains elevation at a constant rate.
After 2 miles, the elevation is 5,500 feet above sea level.
After 4 miles, the elevation is 6,200 feet above sea level. (Lesson 3-6)

   a. Find the elevation of the road at the point where the road begins.

   b. Describe where you would see the point in part (a) on a graph where y represents the elevation in feet and x represents the distance along the road in miles.

5. Match each graph to a situation. **(Lesson 3-6)**

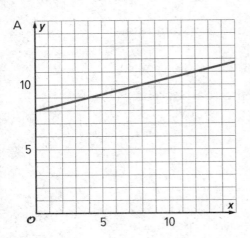

A

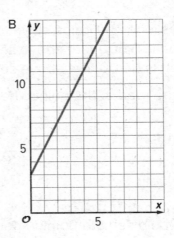

B

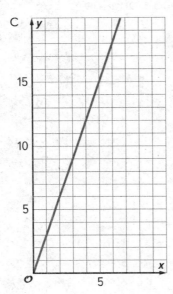

C

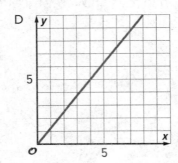

D

a. The graph represents the perimeter, y, in units, for an equilateral triangle with side length of x units. The slope of the line is 3.

b. The amount of money, y, in a cash box after x tickets are purchased for carnival games. The slope of the line is $\frac{1}{4}$.

c. The number of chapters read, y, after x days. The slope of the line is $\frac{5}{4}$.

d. The graph shows the cost in dollars, y, of a muffin delivery and the number of muffins, x, ordered. The slope of the line is 2.

Lesson 3-9

Slopes Don't Have to be Positive

NAME _____ DATE _____ PERIOD _____

Learning Goal Let's find out what a negative slope means.

Warm Up

9.1 Which One Doesn't Belong: Odd Line Out

Which line doesn't belong?

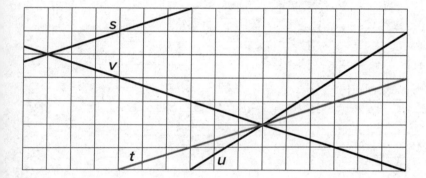

Activity

9.2 Stand Clear of the Closing Doors, Please

Noah put $40 on his fare card. Every time he rides public transportation, $2.50 is subtracted from the amount available on his card.

1. How much money, in dollars, is available on his card after he takes

   a. 0 rides?

   b. 1 ride?

   c. 2 rides?

   d. *x* rides?

2. Graph the relationship between amount of money on the card and number of rides.

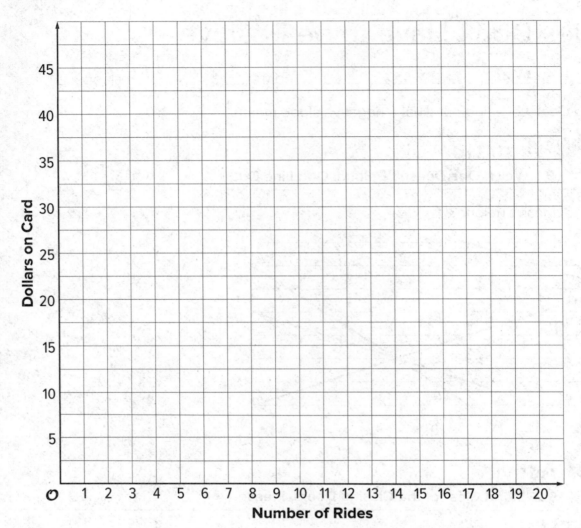

3. How many rides can Noah take before the card runs out of money? Where do you see this number of rides on your graph?

NAME _____ DATE _____ PERIOD _____

Activity
9.3 Travel Habits in July

Here is a graph that shows the amount on Han's fare card for every day of last July.

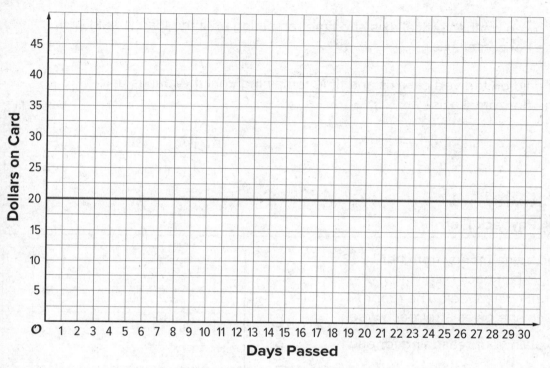

1. Describe what happened with the amount on Han's fare card in July.

2. Plot and label 3 different points on the line.

3. Write an equation that represents the amount on the card in July, *y*, after *x* days.

4. What value makes sense for the slope of the line that represents the amounts on Han's fare card in July?

Let's say you have taken out a loan and are paying it back. Which of the following graphs have positive slope and which have negative slope?

1. Amount paid on the vertical axis and time since payments started on the horizontal axis.

2. Amount owed on the vertical axis and time remaining until the loan is paid off on the horizontal axis.

3. Amount paid on the vertical axis and time remaining until the loan is paid off on the horizontal axis.

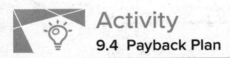

Activity
9.4 Payback Plan

Elena borrowed some money from her brother. She pays him back by giving him the same amount every week. The graph shows how much she owes after each week.

Answer and explain your reasoning for each question.

1. What is the slope of the line?

2. Explain how you know whether the slope is positive or negative.

3. What does the slope represent in this situation?

4. How much did Elena borrow?

5. How much time will it take for Elena to pay back all the money she borrowed?

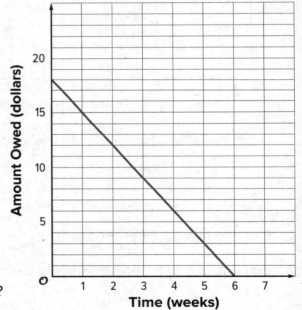

Summary

Slopes Don't Have to be Positive

At the end of winter in Maine, the snow on the ground was 30 inches deep. Then there was a particularly warm day and the snow melted at the rate of 1 inch per hour.

The graph shows the relationship between the time since the snow started to melt and the depth of the snow.

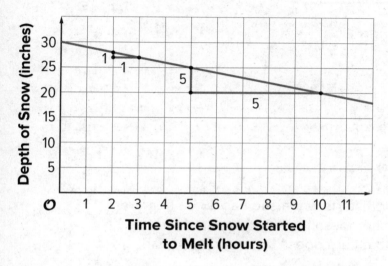

The slope of the graph is -1 since the rate of change is -1 inch per hour. That is, the depth goes *down* 1 inch per hour.

The vertical intercept is 30 since the snow was 30 inches deep when the warmth started to melt the snow.

The two slope triangles show how the rate of change is constant. It just also happens to be negative in this case since after each hour that passes, there is 1 inch *less* snow.

Graphs with negative slope often describe situations where some quantity is decreasing over time, like the depth of snow on warm days or the amount of money on a fare card being used to take rides on buses.

Slopes can be positive, negative, or even zero! A slope of 0 means there is no change in the y-value even though the x-value may be changing.

For example, Elena won a contest where the prize was a special pass that gives her free bus rides for a year. Her fare card had $5 on it when she won the prize.

Here is a graph of the amount of money on her fare card after winning the prize.

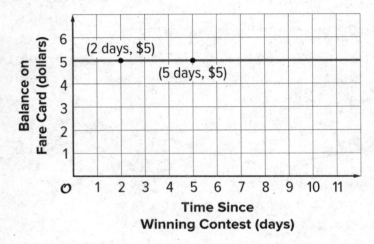

The vertical intercept is 5, since the graph starts when she has $5 on her fare card. The slope of the graph is 0 since she doesn't use her fare card for the next year, meaning the amount on her fare card doesn't change for a year.

In fact, all graphs of linear relationships with slopes equal to 0 are horizontal—a rate of change of 0 means that, from one point to the next, the y-values remain the same.

NAME _____ DATE _____ PERIOD _____

Practice
Slopes Don't Have to be Positive

1. Suppose that during its flight, the elevation e (in feet) of a certain airplane and its time, t, in minutes since takeoff, are related by a linear equation. Consider the graph of this equation, with time represented on the horizontal axis and elevation on the vertical axis. For each situation, decide if the slope is positive, zero, or negative.

   a. The plane is cruising at an altitude of 37,000 feet above sea level.

   b. The plane is descending at rate of 1,000 feet per minute.

   c. The plane is ascending at a rate of 2,000 feet per minute.

2. A group of hikers park their car at a trail head and walk into the forest to a campsite. The next morning, they head out on a hike from their campsite walking at a steady rate. The graph shows their distance in miles, d, from the car after h hours of hiking.

   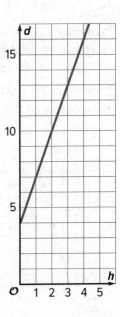

   a. How far is the campsite from their car? Explain how you know.

   b. Write an equation that describes the relationship between d and h.

   c. After how many hours of hiking will they be 16 miles from their car? Explain or show your reasoning.

3. Elena's aunt pays her $1 for each call she makes to let people know about her aunt's new business. The table shows how much money Diego receives for washing windows for his neighbors. (Lesson 3-4)

Number of Windows	Number of Dollars
27	30
45	50
81	90

Select **all** the statements about the situation that are true.

A. Elena makes more money for making 10 calls than Diego makes for washing 10 windows.

B. Diego makes more money for washing each window than Elena makes for making each call.

C. Elena makes the same amount of money for 20 calls as Diego makes for 18 windows.

D. Diego needs to wash 35 windows to make as much money as Elena makes for 40 calls.

E. The equation $y = \frac{9}{10}x$, where y is number of dollars and x is number of windows, represents Diego's situation.

F. The equation $y = x$, where y is the number of dollars and x is the number of calls, represents Elena's situation.

4. Each square on a grid represents 1 unit on each side. Match the graphs with the slopes of the lines.

A

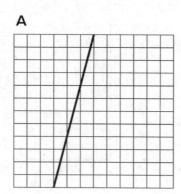

B

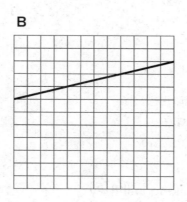

C

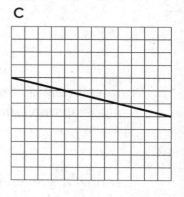

a. $-\frac{1}{4}$

b. $\frac{1}{4}$

c. 4

Volume 1

McGraw-Hill
Illustrative Mathematics™
Course 3

Mc
Graw
Hill

Cover: (l)Liyao Xie/Moment/Getty Images, (tr)Pobytov/DigitalVision Vectors/
Getty Images, (cr)Nikita Veremcuks/EyeEm/Getty Images, (br)MirageC/
Moment/Getty Images

mheducation.com/prek-12

Send all inquiries to:
McGraw-Hill Education
STEM Learning Solutions Center
8787 Orion Place
Columbus, OH 43240

ISBN: 978-0-07-688463-6
MHID: 0-07-688463-5

Illustrative Mathematics, Course 3
Student Edition, Volume 1

Printed in the United States of America.

10 11 12 LMN 28 27 26 25 24 23 22

'Notice and Wonder' and 'I Notice/I Wonder' are trademarks of the National
Council of Teachers of Mathematics, reflecting approaches developed by the
Math Forum (http://www.nctm.org/mathforum/), and used here with
permission.

Lesson 3-10

Calculating Slope

NAME _____ DATE _____ PERIOD _____

Learning Goal Let's calculate slope from two points.

Warm Up
10.1 Number Talk: Integer Operations

Find values for a and b that make each side have the same value.

1. $\dfrac{a}{b} = \text{-}2$

2. $\dfrac{a}{b} = 2$

3. $a - b = \text{-}2$

Activity
10.2 Toward a More General Slope Formula

1. Plot the points (1, 11) and (8, 2), and use a ruler to draw the line that passes through them.

2. Without calculating, do you expect the slope of the line through (1, 11) and (8, 2) to be positive or negative? How can you tell?

3. Calculate the slope of this line.

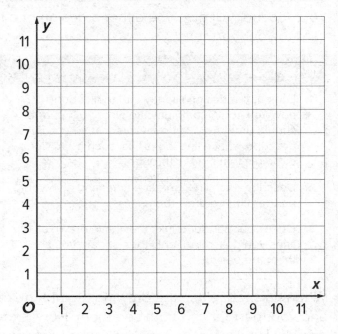

Find the value of k so that the line passing through each pair of points has the given slope.

1. $(k, 2)$ and $(11, 14)$, slope = 2

2. $(1, k)$ and $(4, 1)$, slope = -2

3. $(3, 5)$ and $(k, 9)$, slope = $\frac{1}{2}$

4. $(-1, 4)$ and $(-3, k)$, slope = $\frac{-1}{2}$

5. $\left(\frac{-15}{2}, \frac{3}{16}\right)$ and $\left(\frac{-13}{22}, k\right)$, slope = 0

Activity
10.3 Making Designs

Your teacher will give you either a design or a blank graph. Do not show your card to your partner.

If your teacher gives you the design:	If your teacher gives you the blank graph:
1. Look at the design silently and think about how you could communicate what your partner should draw. Think about ways that you can describe what a line looks like, such as its slope or points that it goes through. 2. Describe each line, one at a time, and give your partner time to draw them. 3. Once your partner thinks they have drawn all the lines you described, only then should you show them the design.	1. Listen carefully as your partner describes each line, and draw each line based on their description. 2. You are not allowed to ask for more information about a line than what your partner tells you. 3. Do not show your drawing to your partner until you have finished drawing all the lines they describe.

When finished, place the drawing next to the card with the design so that you and your partner can both see them. How is the drawing the same as the design? How is it different? Discuss any miscommunication that might have caused the drawing to look different from the design.

Pause here so your teacher can review your work. When your teacher gives you a new set of cards, switch roles for the second problem.

NAME _____ DATE _____ PERIOD _____

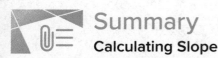

Summary
Calculating Slope

We learned earlier that one way to find the slope of a line is by drawing a slope triangle. For example, using the slope triangle shown here, the slope of the line is $-\frac{2}{4}$, or $-\frac{1}{2}$ (we know the slope is negative because the line is decreasing from left to right).

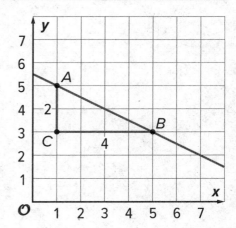

But slope triangles are only one way to calculate the slope of a line. Let's compute the slope of this line a different way using just the points $A = (1, 5)$ and $B = (5, 3)$. Since we know the slope is the vertical change divided by the horizontal change, we can calculate the change in the y-values and then the change in the x-values. Between points A and B, the y-value change is $3 - 5 = -2$ and the x-value change is $5 - 1 = 4$. This means the slope is $-\frac{2}{4}$, or $-\frac{1}{2}$, which is the same as what we found using the slope triangle.

Notice that in each of the calculations, we subtracted the value from point A from the value from point B. If we had done it the other way around, then the y-value change would have been $5 - 3 = 2$ and the x-value change would have been $1 - 5 = -4$, which still gives us a slope of $-\frac{1}{2}$.

But what if we were to mix up the orders? If that had happened, we would think the slope of the line is positive $\frac{1}{2}$ since we would either have calculated $\frac{-2}{-4}$ or $\frac{2}{4}$. Since we already have a graph of the line and can see it has a negative slope, this is clearly incorrect. It we don't have a graph to check our calculation, we could think about how the point on the left, $(1, 5)$, is higher than the point on the right, $(5, 3)$, meaning the slope of the line must be negative.

1. For each graph, calculate the slope of the line.

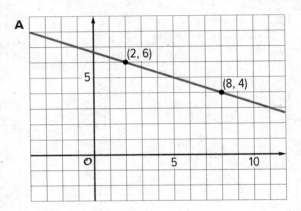

A

(2, 6)

(8, 4)

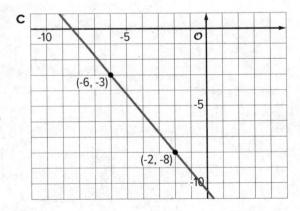

B

(-5, 7)

(1, 1)

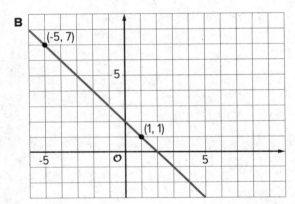

C

(-6, -3)

(-2, -8)

NAME _____ DATE _____ PERIOD _____

2. Match each pair of points to the slope of the line that joins them.

Pair of Points	**Slope**

a. (9, 10) and (7, 2)   4 $\dfrac{-5}{2}$

b. (-8, -11) and (-1, -5)   -3 $\dfrac{6}{7}$

c. (5, -6) and (2, 3)

d. (6, 3) and (5, -1)

e. (4, 7) and (6, 2)

3. Draw a line with the given slope through the given point. What other point lies on that line?

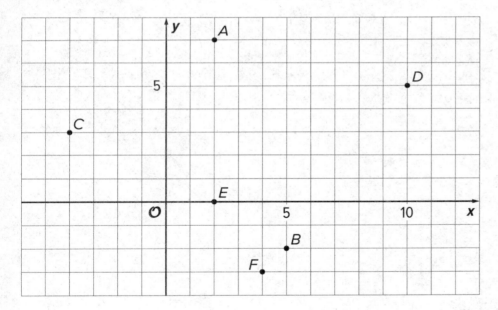

a. Point A, slope = -3

b. Point A, slope = $\dfrac{-1}{4}$

c. Point C, slope = $\dfrac{-1}{2}$

d. Point E, slope = $\dfrac{-2}{3}$

4. Make a sketch of a linear relationship with a slope of 4 and a negative *y*-intercept. Show how you know the slope is 4 and write an equation for the line. **(Lesson 3-8)**

Lesson 3-11

Equations of All Kinds of Lines

NAME _____ DATE _____ PERIOD _____

Learning Goal Let's write equations for vertical and horizontal lines.

Warm Up
11.1 Which One Doesn't Belong: Pairs of Lines

Which one doesn't belong?

Graph A

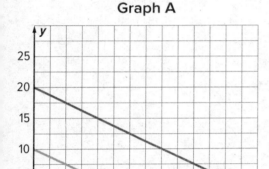

Graph B

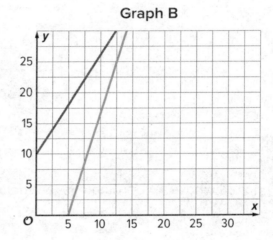

Graph C

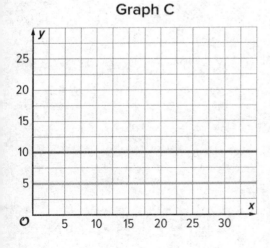

Graph D

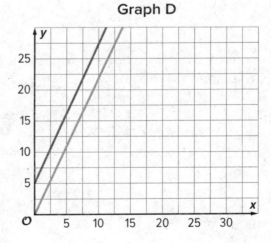

Activity

11.2 All the Same

1. Plot at least 10 points whose y-coordinate is -4. What do you notice about them?

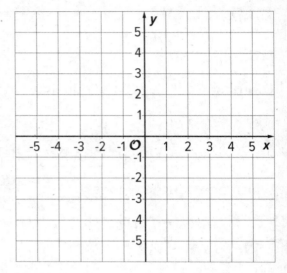

2. Which equation makes the most sense to represent all of the points with y-coordinate -4? Explain how you know.

    $x = -4$

    $y = -4x$

    $y = -4$

    $x + y = -4$

3. Plot at least 10 points whose x-coordinate is 3. What do you notice about them?

4. Which equation makes the most sense to represent all of the points with x-coordinate 3? Explain how you know.

    $x = 3$ $y = 3$

    $y = 3x$ $x + y = 3$

5. Graph the equation $x = -2$.

6. Graph the equation $y = 5$.

NAME _____ DATE _____ PERIOD _____

Are you ready for more?

1. Draw the rectangle with vertices (2, 1), (5, 1), (5, 3), and (2, 3).

2. For each of the four sides of the rectangle, write an equation for a line containing the side.

3. A rectangle has sides on the graphs of $x = -1$, $x = 3$, $y = -1$, and $y = 1$. Find the coordinates of each vertex.

Activity

11.3 Same Perimeter

1. There are many possible rectangles whose perimeter is 50 units. Complete the table with lengths, ℓ, and widths, w, of at least 10 such rectangles.

ℓ										
w										

2. The graph shows one rectangle whose perimeter is 50 units, and has its lower left vertex at the origin and two sides on the axes. On the same graph, draw more rectangles with perimeter 50 units using the values from your table. Make sure that each rectangle has a lower left vertex at the origin and two sides on the axes.

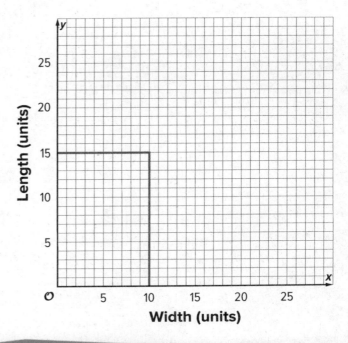

3. Each rectangle has a vertex that lies in the first quadrant. These vertices lie on a line. Draw in this line and write an equation for it.

4. What is the slope of this line? How does the slope describe how the width changes as the length changes (or vice versa)?

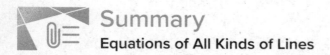

Summary
Equations of All Kinds of Lines

Horizontal lines in the coordinate plane represent situations where the y-value doesn't change at all while the x-value changes. For example, the horizontal line that goes through the point $(0, 13)$ can be described in words as "for all points on the line, the y-value is always 13." An equation that says the same thing is $y = 13$.

Vertical lines represent situations where the x-value doesn't change at all while the y-value changes. The equation $x = -4$ describes a vertical line through the point $(-4, 0)$.

NAME _____ DATE _____ PERIOD _____

Practice
Equations of All Kinds of Lines

1. Suppose you wanted to graph the equation $y = -4x - 1$.

   a. Describe the steps you would take to draw the graph.

   b. How would you check that the graph you drew is correct?

2. Draw the following lines and then write an equation for each.

   a. Slope is 0, y-intercept is 5

   b. Slope is 2, y-intercept is -1

   c. Slope is -2, y-intercept is 1

   d. Slope is $\frac{-1}{2}$, y-intercept is -1

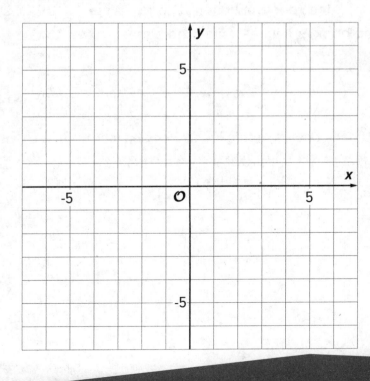

3. Write an equation for each line.

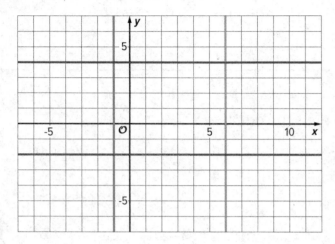

4. A publisher wants to figure out how thick their new book will be. The book has a front cover and a back cover, each of which have a thickness of $\frac{1}{4}$ of an inch. They have a choice of which type of paper to print the book on. **(Lesson 3-7)**

  a. Bond paper has a thickness of $\frac{1}{4}$ inch per one hundred pages. Write an equation for the width of the book, y, if it has x hundred pages, printed on bond paper.

  b. Ledger paper has a thickness of $\frac{2}{5}$ inch per one hundred pages. Write an equation for the width of the book, y, if it has x hundred pages, printed on ledger paper.

  c. If they instead chose front and back covers of thickness $\frac{1}{3}$ of an inch, how would this change the equations in the previous two parts?

Lesson 3-12

Solutions to Linear Equations

NAME _____ DATE _____ PERIOD _____

Learning Goal Let's think about what it means to be a solution to a linear equation with two variables in it.

Warm Up
12.1 Estimate Area

Which figure has the largest shaded region?

A   B   C

Activity

12.2 Apples and Oranges

At the corner produce market, apples cost $1 each and oranges cost $2 each.

1. Find the cost of:

   a. 6 apples and 3 oranges

   b. 4 apples and 4 oranges

   c. 5 apples and 4 oranges

   d. 8 apples and 2 oranges

2. Noah has $10 to spend at the produce market. Can he buy 7 apples and 2 oranges? Explain or show your reasoning.

3. What combinations of apples and oranges can Noah buy if he spends all of his $10?

4. Use two variables to write an equation that represents $10-combinations of apples and oranges. Be sure to say what each variable means.

5. What are three combinations of apples and oranges that make your equation true? What are three combinations of apples and oranges that make it false?

Are you ready for more?

1. Graph the equation you wrote relating the number of apples and the number of oranges.

(continued on the next page)

NAME _____ DATE _____ PERIOD _____

2. What is the slope of the graph? What is the meaning of the slope in terms of the context?

3. Suppose Noah has $20 to spend. Graph the equation describing this situation. What do you notice about the relationship between this graph and the earlier one?

Activity

12.3 Solutions and Everything Else

You have two numbers. If you double the first number and add it to the second number, the sum is 10.

1. Let x represent the first number and let y represent the second number. Write an equation showing the relationship between x, y, and 10.

2. Draw and label a set of x- and y-axes. Plot at least five points on this coordinate plane that make the statement and your equation true. What do you notice about the points you have plotted?

3. List ten points that do *not* make the statement true. Using a different color, plot each point in the same coordinate plane. What do you notice about these points compared to your first set of points?

Think of all the rectangles whose perimeters are 8 units.
If x represents the width and y represents the length, then

$$2x + 2y = 8$$

expresses the relationship between the width and length for all such rectangles.

For example, the width and length could be 1 and 3, since $2 \cdot 1 + 2 \cdot 3 = 8$ or the width and length could be 2.75 and 1.25, since $2 \cdot (2.75) + 2 \cdot (1.25) = 8$.

We could find many other possible pairs of width and length, (x, y), that make the equation true—that is, pairs (x, y) that when substituted into the equation make the left side and the right side equal.

A **solution to an equation with two variables** is any pair of values (x, y) that make the equation true.

We can think of the pairs of numbers that are solutions of an equation as points on the coordinate plane.

Here is a line created by all the points (x, y) that are solutions to $2x + 2y = 8$.

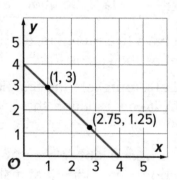

• Every point on the line represents a rectangle whose perimeter is 8 units.

• All points not on the line are not solutions to $2x + 2y = 8$.

> **Glossary**
>
> **solution to an equation with two variables**

NAME _____ DATE _____ PERIOD _____

Practice
Solutions to Linear Equations

1. Select **all** of the ordered pairs (x, y) that are solutions to the linear equation $2x + 3y = 6$.

   (A.) $(0, 2)$

   (B.) $(0, 6)$

   (C.) $(2, 3)$

   (D.) $(3, -2)$

   (E.) $(3, 0)$

   (F.) $(6, -2)$

2. The graph shows a linear relationship between x and y. x represents the number of comic books Priya buys at the store, all at the same price, and y represents the amount of money (in dollars) Priya has after buying the comic books.

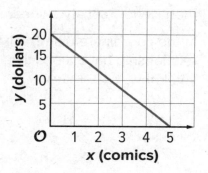

   a. Find and interpret the x- and y-intercepts of this line.

   b. Find and interpret the slope of this line.

   c. Find an equation for this line.

   d. If Priya buys 3 comics, how much money will she have remaining?

3. Match each equation with its three solutions.

Equation	Solutions
a. $y = 1.5x$	i. $(14, 21), (2, 3), (8, 12)$
b. $2x + 3y = 7$	ii. $(-3, -7), (0, -4), (-1, -5)$
c. $x - y = 4$	iii. $\left(\frac{1}{8}, \frac{7}{8}\right), \left(\frac{1}{2}, \frac{1}{2}\right), \left(\frac{1}{4}, \frac{3}{4}\right)$
d. $3x = \frac{y}{2}$	iv. $\left(1, 1\frac{2}{3}\right), (-1, 3), \left(0, 2\frac{1}{3}\right)$
e. $y = -x + 1$	v. $(0.5, 3), (1, 6), (1.2, 7.2)$

4. A container of fuel dispenses fuel at the rate of 5 gallons per second. If y represents the amount of fuel remaining in the container, and x represents the number of seconds that have passed since the fuel started dispensing, then x and y satisfy a linear relationship.

   In the coordinate plane, will the slope of the line representing that relationship have a positive, negative, or zero slope? Explain how you know. (Lesson 3-10)

5. A sandwich store charges a delivery fee to bring lunch to an office building. One office pays $33 for 4 turkey sandwiches. Another office pays $61 for 8 turkey sandwiches. How much does each turkey sandwich add to the cost of the delivery? Explain how you know. (Lesson 3-5)

Lesson 3-13

More Solutions to Linear Equations

NAME _____ DATE _____ PERIOD _____

Learning Goal Let's find solutions to more linear equations.

 ## Warm Up
13.1 Coordinate Pairs

For each equation choose a value for x and then solve to find the corresponding y-value that makes that equation true.

1. $6x = 7y$ **2.** $5x + 3y = 9$ **3.** $y + 5 - \frac{1}{3}x = 7$

 ## Activity
13.2 True or False: Solutions in the Coordinate Plane

Here are graphs representing three linear relationships. These relationships could also be represented with equations.

For each statement below, decide if it is true or false. Explain your reasoning.

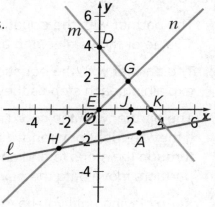

1. (4, 0) is a solution of the equation for line m.

2. The coordinates of the point G make both the equation for line m and the equation for line n true.

3. $x = 0$ is a solution of the equation for line n.

4. (2, 0) makes both the equation for line m and the equation for line n true.

5. There is no solution for the equation for line ℓ that has $y = 0$.

6. The coordinates of point H are solutions to the equation for line ℓ.

7. There are exactly two solutions of the equation for line ℓ.

8. There is a point whose coordinates make the equations of all three lines true.

After you finish discussing the eight statements, find another group and check your answers against theirs. Discuss any disagreements.

 Activity

13.3 I'll Take an X, Please

One partner has 6 cards labeled A through F and one partner has 6 cards labeled a through f. In each pair of cards (for example, Cards A and a), there is an equation on one card and a coordinate pair, (x, y), that makes the equation true on the other card.

1. The partner with the equation asks the partner with a solution for either the x-value or the y-value and explains why they chose the one they did.

2. The partner with the equation uses this value to find the other value, explaining each step as they go.

3. The partner with the coordinate pair then tells the partner with the equation if they are right or wrong. If they are wrong, both partners should look through the steps to find and correct any errors. If they are right, both partners move onto the next set of cards.

4. Keep playing until you have finished Cards A through F.

NAME _____ DATE _____ PERIOD _____

Are you ready for more?

Consider the equation $ax + by = c$, where a, b, and c are positive numbers.

1. Find the coordinates of the x- and y-intercepts of the graph of the equation.

2. Find the slope of the graph.

Summary
More Solutions to Linear Equations

Let's think about the linear equation $2x - 4y = 12$. If we know $(0, -3)$ is a solution to the equation, then we also know $(0, -3)$ is a point on the graph of the equation. Since this point is on the y-axis, we also know that it is the vertical intercept of the graph.

But what about the coordinate of the horizontal intercept, when $y = 0$? Well, we can use the equation to figure it out.

$$2x - 4y = 12$$
$$2x - 4(0) = 12$$
$$2x = 12$$
$$x = 6$$

Since $x = 6$ when $y = 0$, we know the point $(6, 0)$ is on the graph of the line.

No matter the form a linear equation comes in, we can always find solutions to the equation by starting with one value and then solving for the other value.

Practice

More Solutions to Linear Equations

1. For each equation, find y when $x = -3$. Then find x when $y = 2$.

   a. $y = 6x + 8$

   b. $y = \frac{2}{3}x$

   c. $y = -x + 5$

   d. $y = \frac{3}{4}x - 2\frac{1}{2}$

   e. $y = 1.5x + 11$

2. True or false: The points (6, 13), (21, 33), and (99, 137) all lie on the same line. The equation of the line is $y = \frac{4}{3}x + 5$. Explain or show your reasoning.

NAME _____ DATE _____ PERIOD _____

3. Here is a linear equation: $y = \frac{1}{4}x + \frac{5}{4}$

a. Are (1, 1.5) and (12, 4) solutions to the equation? Explain or show your reasoning.

b. Find the x-intercept of the graph of the equation. Explain or show your reasoning.

4. Find the coordinates of *B*, *C*, and *D* given that $AB = 5$ and $BC = 10$. **(Lesson 2-11)**

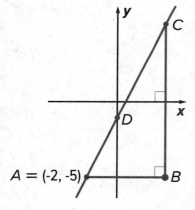

5. Match each graph of a linear relationship to a situation that most reasonably reflects its context. **(Lesson 3-9)**

A

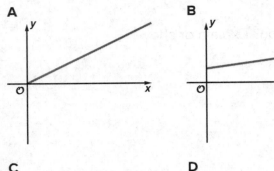

B

C

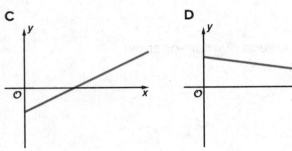

D

a. y is the weight of a kitten x days after birth.

b. y is the distance left to go in a car ride after x hours of driving at a constant rate toward its destination.

c. y is the temperature, in degrees C, of a gas being warmed in a laboratory experiment.

d. y is the amount of calories consumed eating x crackers.

Lesson 3-14

Using Linear Relations to Solve Problems

NAME _____ DATE _____ PERIOD _____

Learning Goal Let's write equations for real-world situations and think about their solutions.

 ## Warm Up
14.1 Buying Fruit

For each relationship described, write an equation to represent the relationship.

1. Grapes cost $2.39 per pound and bananas cost $0.59 per pound. You have $15 to spend on g pounds of grapes and b pounds of bananas.

2. A savings account has $50 in it at the start of the year and $20 is deposited each week. After x weeks, there are y dollars in the account.

Activity

14.2 Five Savings Accounts

Each line represents one person's weekly savings account balance from the start of the year.

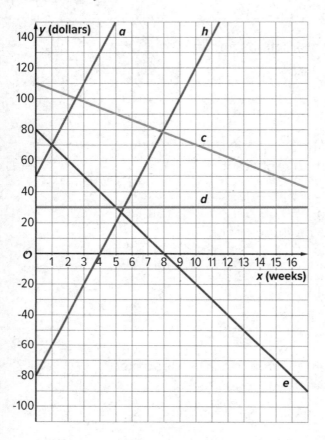

1. Choose one line and write a description of what happens to that person's account over the first 17 weeks of the year. Do not tell your group which line you chose.

2. Share your story with your group and see if anyone can guess your line.

3. Write an equation for each line on the graph. What do the slope, *m*, and vertical intercept, *b*, in each equation mean in the situation?

NAME _____ DATE _____ PERIOD _____

4. For which equation is (1, 70) a solution? Interpret this solution in terms of your story.

5. Predict the balance in each account after 20 weeks.

Activity

14.3 Fabulous Fish

The Fabulous Fish Market orders tilapia, which costs $3 per pound, and salmon, which costs $5 per pound. The market budgets $210 to spend on this order each day.

1. What are five different combinations of salmon and tilapia that the market can order?

2. Define variables and write an equation representing the relationship between the amount of each fish bought and how much the market spends.

3. Sketch a graph of the relationship. Label your axes.

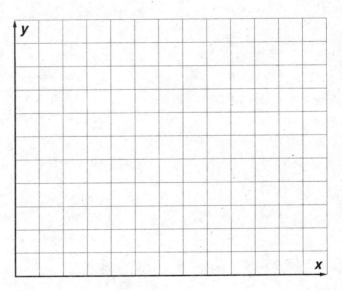

4. On your graph, plot and label the combinations A—F.

	A	B	C	D	E	F
Pounds of Tilapia	5	19	27	25	65	55
Pounds of Salmon	36	30.6	25	27	6	4

5. Which of these combinations can the market order?
Explain or show your reasoning.

6. List two ways you can tell if a pair of numbers is a solution to an equation.

Learning Targets

Lesson	Learning Target(s)
3-1 Understanding Proportional Relationships	• I can graph a proportional relationship from a story. • I can use the constant of proportionality to compare the pace of different animals.
3-2 Graphs of Proportional Relationships	• I can graph a proportional relationship from an equation. • I can tell when two graphs are of the same proportional relationship even if the scales are different.
3-3 Representing Proportional Relationships	• I can scale and label coordinate axes in order to graph a proportional relationship.
3-4 Comparing Proportional Relationships	• I can compare proportional relationships represented in different ways.

(continued on the next page)

(continued from the previous page)

Lesson	Learning Target(s)
3-5 Introduction to Linear Relationships	• I can find the rate of change of a linear relationship by figuring out the slope of the line representing the relationship.
3-6 More Linear Relationships	• I can interpret the vertical intercept of a graph of a real-world situation. • I can match graphs to the real-world situations they represent by identifying the slope and the vertical intercept.
3-7 Representations of Linear Relationships	• I can use patterns to write a linear equation to represent a situation. • I can write an equation for the relationship between the total volume in a graduated cylinder and the number of objects added to the graduated cylinder.
3-8 Translating to $y = mx + b$	• I can explain where to find the slope and vertical intercept in both an equation and its graph. • I can write equations of lines using $y = mx + b$.

Lesson		Learning Target(s)
3-9	Slopes Don't Have to be Positive	• I can give an example of a situation that would have a negative slope when graphed. • I can look at a graph and tell if the slope is positive or negative and explain how I know.
3-10	Calculating Slope	• I can calculate positive and negative slopes given two points on the line. • I can describe a line precisely enough that another student can draw it.
3-11	Equations of All Kinds of Lines	• I can write equations of lines that have a positive or a negative slope. • I can write equations of vertical and horizontal lines.
3-12	Solutions to Linear Equations	• I know that the graph of an equation is a visual representation of all the solutions to the equation. • I understand what the solution to an equation in two variables is.

(continued on the next page)

(continued from the previous page)

Lesson	Learning Target(s)
3-13 More Solutions to Linear Equations	• I can find solutions (x, y) to linear equations given either the x- or the y-value to start from.
3-14 Using Linear Relations to Solve Problems	• I can write linear equations to reason about real-world situations.

Notes:

Linear Equations and Linear Systems

At the end of this unit, you'll apply what you learned about linear systems to solve problems about cycling at different rates.

Topics

- Puzzle Problems
- Linear Equations in One Variable
- Systems of Linear Equations
- Let's Put It to Work

Linear Equations and Linear Systems

Lesson 4-1
Number Puzzles

NAME _____ DATE _____ PERIOD _____

Learning Goal Let's solve some puzzles!

Warm Up
1.1 Notice and Wonder: A Number Line

What do you notice? What do you wonder?

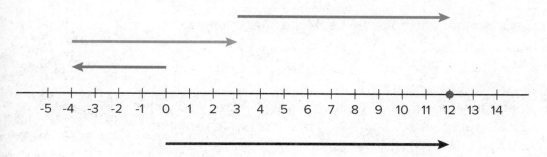

Activity
1.2 Telling Temperatures

Solve each puzzle. Show your thinking. Organize it so it can be followed by others.

1. The temperature was very cold. Then the temperature doubled. Then the temperature dropped by 10 degrees. Then the temperature increased by 40 degrees. The temperature is now 16 degrees. What was the starting temperature?

2. Lin ran twice as far as Diego. Diego ran 300 m farther than Jada. Jada ran $\frac{1}{3}$ the distance that Noah ran. Noah ran 1200 m. How far did Lin run?

Activity
1.3 Making a Puzzle

Write another number puzzle with at least three steps. On a different piece of paper, write a solution to your puzzle.

Trade puzzles with your partner and solve theirs. Make sure to show your thinking.

With your partner, compare your solutions to each puzzle. Did they solve them the same way you did? Be prepared to share with the class which solution strategy you like best.

Are you ready for more?

Here is a number puzzle that uses math. Some might call it a magic trick!

1. Think of a number.
2. Double the number.
3. Add 9.
4. Subtract 3.
5. Divide by 2.
6. Subtract the number you started with.
7. The answer should be 3.

Why does this always work? Can you think of a different number puzzle that uses math (like this one) that will always result in 5?

NAME _____ DATE _____ PERIOD _____

Summary
Number Puzzles

Here is an example of a puzzle problem:

Twice a number plus 4 is 18. What is the number?

There are many different ways to represent and solve puzzle problems.

- We can reason through it.

  Twice a number plus 4 is 18.

  Then twice the number is $18 - 4 = 14$.

  That means the number is 7.

- We can draw a diagram.

x	x	4

18

x	x

14

x

7

- We can write and solve an equation.

$$2x + 4 = 18$$
$$2x = 14$$
$$x = 7$$

Reasoning and diagrams help us see what is going on and why the answer is what it is. But as number puzzles and story problems get more complex, those methods get harder, and equations get more and more helpful. We will use different kinds of diagrams to help us understand problems and strategies in future lessons, but we will also see the power of writing and solving equations to answer increasingly more complex mathematical problems.

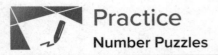

Practice
Number Puzzles

1. Tyler reads $\frac{2}{15}$ of a book on Monday, $\frac{1}{3}$ of it on Tuesday, $\frac{2}{9}$ of it on Wednesday, and $\frac{3}{4}$ of the remainder on Thursday. If he still has 14 pages left to read on Friday, how many pages are there in the book?

2. Clare asks Andre to play the following number puzzle:

   - Pick a number
   - Add 2
   - Multiply by 3
   - Subtract 7
   - Add your original number

   Andre's final result is 27. Which number did he start with?

3. In a basketball game, Elena scores twice as many points as Tyler. Tyler scores four points fewer than Noah, and Noah scores three times as many points as Mai. If Mai scores 5 points, how many points did Elena score? Explain your reasoning.

4. Select **all** of the given points in the coordinate plane that lie on the graph of the linear equation $4x - y = 3$. (Lesson 3-12)

   (A.) (-1, -7)   (D.) (1, 1)

   (B.) (0, 3)   (E.) (2, 5)

   (C.) ($\frac{3}{4}$, 0)   (F.) (4, -1)

NAME _____ DATE _____ PERIOD _____

5. A store is designing the space for rows of nested shopping carts. Each row
has a starting cart that is 4 feet long, followed by the nested carts
(so 0 nested carts means there's just the starting cart). The store measured
a row of 13 nested carts to be 23.5 feet long, and a row of 18 nested carts
to be 31 feet long. **(Lesson 3-5)**

a. Create a graph of the situation.

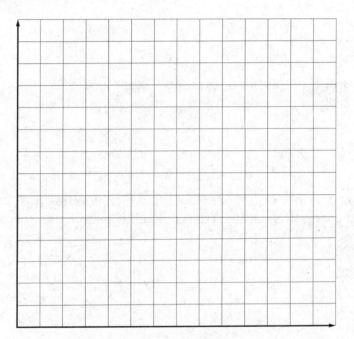

b. How much does each nested cart add to the length of the row?
Explain your reasoning.

c. If the store design allows for 43 feet for each row, how many total carts
fit in a row?

6. Triangle *A* is an isosceles triangle with two angles of measure *x* degrees and one angle of measure *y* degrees. (Lesson 3-13)

   a. Find three combinations of *x* and *y* that make this sentence true.

   b. Write an equation relating *x* and *y*.

   c. If you were to sketch the graph of this linear equation, what would its slope be? How can you interpret the slope in the context of the triangle?

7. Consider the following graphs of linear equations. Decide which line has a positive slope, and which has a negative slope. Then calculate each line's exact slope. (Lesson 3-10)

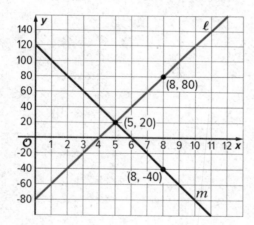

Lesson 4-2

Keeping the Equation Balanced

NAME _____ DATE _____ PERIOD _____

Learning Goal Let's figure out unknown weights on balanced hangers.

Warm Up
2.1 Notice and Wonder: Hanging Socks

What do you notice? What do you wonder?

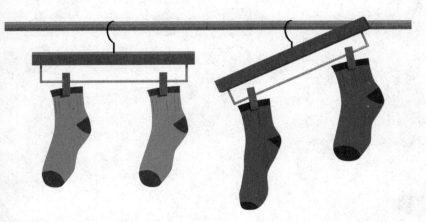

Activity
2.2 Hanging Blocks

This picture represents a hanger that is balanced because the weight on each side is the same.

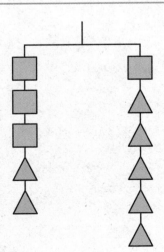

1. Elena takes two triangles off of the left side and three triangles off of the right side. Will the hanger still be in balance, or will it tip to one side? Which side? Explain how you know.

2. If a triangle weighs 1 gram, how much does a square weigh?

A triangle weighs 3 grams and a circle weighs 6 grams.

1. Find the weight of a square in Hanger A and the weight of a pentagon in Hanger B.

2. Write an equation to represent each hanger.

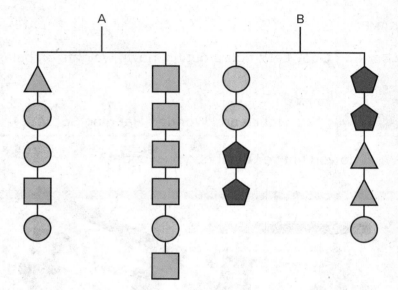

Are you ready for more?

What is the weight of a square on this hanger if a triangle weighs 3 grams?

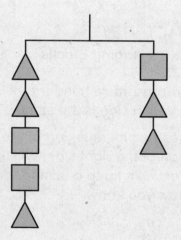

NAME _____ DATE _____ PERIOD _____

Summary
Keeping the Equation Balanced

If we have equal weights on the ends of a hanger, then the hanger will be in balance. If there is more weight on one side than the other, the hanger will tilt to the heavier side.

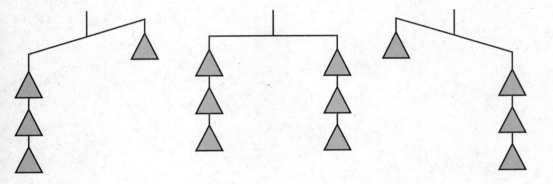

We can think of a balanced hanger as a metaphor for an equation. An equation says that the expressions on each side have equal value, just like a balanced hanger has equal weights on each side.

If we have a balanced hanger and add or remove the same amount of weight from each side, the result will still be in balance.

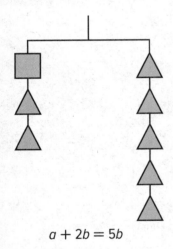

$a + 2b = 5b$

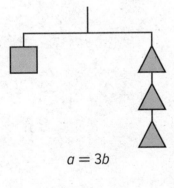

$a = 3b$

We can do these moves with equations as well: adding or subtracting the same amount from each side of an equation maintains the equality.

Practice

Keeping the Equation Balanced

1. Which of the changes would keep the hanger in balance? Select all that apply.

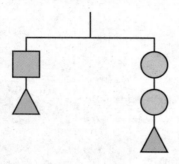

(A.) adding two circles on the left and a square on the right

(B.) adding 2 triangles to each side

(C.) adding two circles on the right and a square on the left

(D.) adding a circle on the left and a square on the right

(E.) adding a triangle on the left and a square on the right

NAME _____ DATE _____ PERIOD _____

2. Here is a balanced hanger diagram.

Each triangle weighs 2.5 pounds, each circle weighs 3 pounds, and x represents the weight of each square. Select *all* equations that represent the hanger.

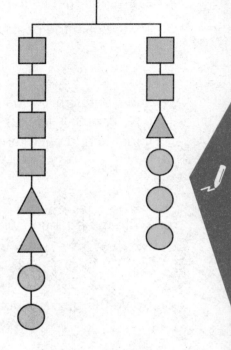

(A.) $x + x + x + x + 11 = x + 11.5$

(B.) $2x = 0.5$

(C.) $4x + 5 + 6 = 2x + 2.5 + 6$

(D.) $2x + 2.5 = 3$

(E.) $4x + 2.5 + 2.5 + 3 + 3 = 2x + 2.5 + 3 + 3 + 3$

3. What is the weight of a square if a triangle weighs 4 grams? Explain your reasoning.

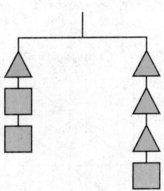

4. Andre came up with the following puzzle. "I am three years younger than my brother, and I am 2 years older than my sister. My mom's age is one less than three times my brother's age. When you add all our ages, you get 87. What are our ages?" (Lesson 4-1)

   a. Try to solve the puzzle.

   b. Jada writes this equation for the sum of the ages:
      $(x) + (x + 3) + (x - 2) + 3(x + 3) - 1 = 87$. Explain the meaning of the variable and each term of the equation.

   c. Write the equation with fewer terms.

   d. Solve the puzzle if you haven't already.

5. These two lines are parallel. Write an equation for each. (Lesson 3-8)

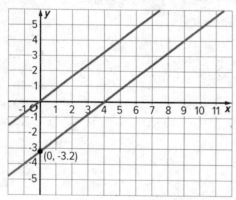

(0, -3.2)

Lesson 4-3

Balanced Moves

NAME _____ DATE _____ PERIOD _____

Learning Goal Let's rewrite equations while keeping the same solutions.

Warm Up
3.1 Matching Hangers

Figures A, B, C, and D show the result of simplifying the hanger in Figure A by removing equal weights from each side.

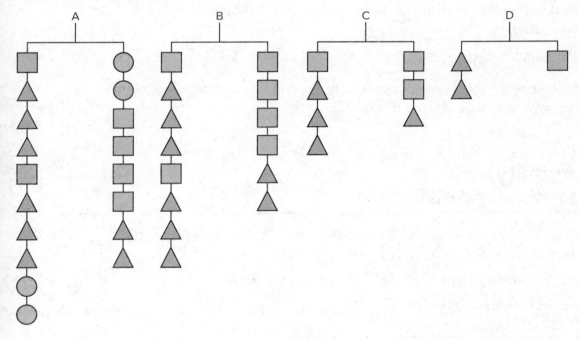

Here are some equations. Each equation represents one of the hanger diagrams.

$2(x + 3y) = 4x + 2y$ $2y = x$

$2(x + 3y) + 2z = 2z + 4x + 2y$ $x + 3y = 2x + y$

1. Write the equation that goes with each figure:

   A: B:

   C: D:

2. Each variable (x, y, and z) represents the weight of one shape. Which goes with which?

3. Explain what was done to each equation to create the next equation. If you get stuck, think about how the hangers changed.

Activity
3.2 Matching Equation Moves

Your teacher will give you some cards. Each of the cards 1 through 6 show two equations. Each of the cards A through E describe a move that turns one equation into another.

1. Match each number card with a letter card.

2. One of the letter cards will not have a match. For this card, write two equations showing the described move.

Activity
3.3 Keeping Equality

1. Noah and Lin both solved the equation $14a = 2(a - 3)$. Do you agree with either of them? Why?

Noah's solution:

$$14a = 2(a - 3)$$
$$14a = 2a - 6$$
$$12a = \text{-}6$$
$$a = -\frac{1}{2}$$

Lin's solution:

$$14a = 2(a - 3)$$
$$7a = a - 3$$
$$6a = \text{-}3$$
$$a = -\frac{1}{2}$$

2. Elena is asked to solve $15 - 10x = 5(x + 9)$. What do you recommend she does to each side first?

3. Diego is asked to solve $3x - 8 = 4(x + 5)$. What do you recommend he does to each side first?

NAME _____ DATE _____ PERIOD _____

Are you ready for more?

In a cryptarithmetic puzzle, the digits 0–9 are represented with letters of the alphabet. Use your understanding of addition to find which digits go with the letters A, B, E, G, H, L, N, and R.

HANGER + HANGER + HANGER = ALGEBRA

Summary

Balanced Moves

An equation tells us that two expressions have equal value. For example, if $4x + 9$ and $-2x - 3$ have equal value, we can write the equation . . .

$$4x + 9 = -2x - 3$$

Earlier, we used hangers to understand that if we add the same positive number to each side of the equation, the sides will still have equal value. It also works if we add *negative numbers*!
For example, we can add -9 to each side of the equation.

$$4x + 9 + \text{-}9 = \text{-}2x - 3 + \text{-}9 \qquad \text{Add -9 to each side.}$$
$$4x = \text{-}2x - 12 \qquad\qquad\quad \text{Combine like terms.}$$

Since expressions represent numbers, we can also add *expressions* to each side of an equation.
For example, we can add $2x$ to each side and still maintain equality.

$$4x + 2x = \text{-}2x - 12 + 2x \qquad \text{Add } 2x \text{ to each side.}$$
$$6x = \text{-}12 \qquad\qquad\qquad\quad \text{Combine like terms.}$$

If we multiply or divide the expressions on each side of an equation by the same number, we will also maintain the equality (so long as we do not divide by zero).

$$6x \cdot \frac{1}{6} = \text{-}12 \cdot \frac{1}{6} \qquad \text{Multiply each side by } \frac{1}{6}.$$

or

$$6x \div 6 = \text{-}12 \div 6 \qquad \text{Divide each side by 6.}$$

Now we can see that $x = \text{-}2$ is the solution to our equation.

We will use these moves in systematic ways to solve equations in future lessons.

Practice
Balanced Moves

1. In this hanger, the weight of the triangle is x and the weight of the square is y.

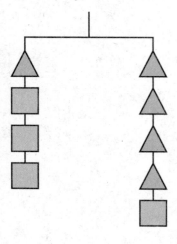

   a. Write an equation using x and y to represent the hanger.

   b. If x is 6, what is y?

2. Andre and Diego were each trying to solve $2x + 6 = 3x - 8$. Describe the first step they each make to the equation.

   a. The result of Andre's first step was $-x + 6 = -8$.

   b. The result of Diego's first step was $6 = x - 8$.

NAME _____ DATE _____ PERIOD _____

3. Respond to each of the following. **(Lesson 3-11)**

   a. Complete the table with values for *x* or *y* that make this equation true:
   $3x + y = 15$.

x	2		6	0	3		
y		3				0	8

   b. Create a graph, plot these points, and find the slope of the line that
   goes through them.

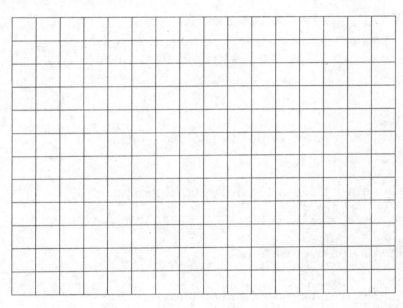

4. Match each set of equations with the move that turned the first equation into the second.

a. $6x + 9 = 4x - 3$
   $2x + 9 = -3$

i. Multiply both sides by $\frac{-1}{4}$.

b. $-4(5x - 7) = -18$
   $5x - 7 = 4.5$

ii. Multiply both sides by -4.

c. $8 - 10x = 7 + 5x$
   $4 - 10x = 3 + 5x$

iii. Multiply both sides by $\frac{1}{4}$.

d. $\dfrac{-5x}{4} = 4$
   $5x = -16$

iv. Add -4x to both sides.

e. $12x + 4 = 20x + 24$
   $3x + 1 = 5x + 6$

v. Add -4 to both sides.

5. Select **all** the situations for which only zero or positive solutions make sense. (Lesson 3-14)

A. Measuring temperature in degrees Celsius at an Arctic outpost each day in January.

B. The height of a candle as it burns over an hour.

C. The elevation above sea level of a hiker descending into a canyon.

D. The number of students remaining in school after 6:00 p.m.

E. A bank account balance over a year.

F. The temperature in degrees Fahrenheit of an oven used on a hot summer day.

Lesson 4-4

More Balanced Moves

NAME _____ DATE _____ PERIOD _____

Learning Goal Let's rewrite some more equations while keeping the same solutions.

 Warm Up
4.1 Different Equations?

Equation 1

$$x - 3 = 2 - 4x$$

Which of these have the same solution as Equation 1? Be prepared to explain your reasoning.

Equation A

$2x - 6 = 4 - 8x$

Equation B

$x - 5 = \text{-}4x$

Equation C

$2(1 - 2x) = x - 3$

Equation D

$\text{-}3 = 2 - 5x$

4.2 Step by Step by Step by Step

Here is an equation, and then all the steps Clare wrote to solve it:

$$14x - 2x + 3 = 3(5x + 9)$$
$$12x + 3 = 3(5x + 9)$$
$$3(4x + 1) = 3(5x + 9)$$
$$4x + 1 = 5x + 9$$
$$1 = x + 9$$
$$-8 = x$$

Here is the same equation, and the steps Lin wrote to solve it:

$$14x - 2x + 3 = 3(5x + 9)$$
$$12x + 3 = 3(5x + 9)$$
$$12x + 3 = 15x + 27$$
$$12x = 15x + 24$$
$$-3x = 24$$
$$x = -8$$

1. Are both of their *solutions* correct? Explain your reasoning.

2. Describe some ways the steps they took are alike and different.

3. Mai and Noah also solved the equation, but some of their steps have errors. Find the incorrect step in each solution and explain why it is incorrect.

Mai:
$$14x - 2x + 3 = 3(5x + 9)$$
$$12x + 3 = 3(5x + 9)$$
$$7x + 3 = 3(9)$$
$$7x + 3 = 27$$
$$7x = 24$$
$$x = \frac{24}{7}$$

Noah:
$$14x - 2x + 3 = 3(5x + 9)$$
$$12x + 3 = 15x + 27$$
$$27x + 3 = 27$$
$$27x = 24$$
$$x = \frac{24}{27}$$

NAME _____ DATE _____ PERIOD _____

Activity
4.3 Make Your Own Steps

Solve these equations for *x*.

1. $\dfrac{12 + 6x}{3} = \dfrac{5 - 9}{2}$

2. $x - 4 = \dfrac{1}{3}(6x - 54)$

3. $-(3x - 12) = 9x - 4$

Are you ready for more?

I have 24 pencils and 3 cups. The second cup holds one more pencil than the first. The third holds one more than the second. How many pencils does each cup contain?

How do we make sure the solution we find for an equation is correct? Accidentally adding when we meant to subtract, missing a negative when we distribute, forgetting to write an x from one line to the next–there are many possible mistakes to watch out for!

Fortunately, each step we take solving an equation results in a new equation with the same solution as the original. This means we can check our work by substituting the value of the solution into the original equation. For example, say we solve the following equation:

$$2x = -3(x + 5)$$
$$2x = -3x + 15$$
$$5x = 15$$
$$x = 3$$

Substituting 3 in place of x into the original equation,

$$2(3) = -3(3 + 5)$$
$$6 = -3(8)$$
$$6 = -24$$

we get a statement that isn't true! This tells us we must have made a mistake somewhere. Checking our original steps carefully, we made a mistake when distributing -3. Fixing it, we now have

$$2x = -3(x + 5)$$
$$2x = -3x - 15$$
$$5x = -15$$
$$x = -3$$

Substituting -3 in place of x into the original equation to make sure we didn't make another mistake:

$$2(-3) = -3(-3 + 5)$$
$$-6 = -3(2)$$
$$-6 = -6$$

This equation is true, so $x = -3$ is the solution.

NAME _____ DATE _____ PERIOD _____

Practice
More Balanced Moves

1. Mai and Tyler work on the equation $\frac{2}{5}b + 1 = -11$ together. Mai's solution is $b = -25$ and Tyler's is $b = -28$. Here is their work:

Mai:

$\frac{2}{5}b + 1 = -11$

$\frac{2}{5}b = -10$

$b = -10 \cdot \frac{5}{2}$

$b = -25$

Tyler:

$\frac{2}{5}b + 1 = -11$

$2b + 1 = -55$

$2b = -56$

$b = -28$

Do you agree with their solutions? Explain or show your reasoning.

2. Solve $3(x - 4) = 12x$.

3. Describe what is being done in each step while solving the equation.

   a. $2(-3x + 4) = 5x + 2$

   b. $-6x + 8 = 5x + 2$

   c. $8 = 11x + 2$

   d. $6 = 11x$

   e. $x = \frac{6}{11}$

4. Andre solved an equation, but when he checked his answer he saw his solution was incorrect. He knows he made a mistake, but he can't find it. Where is Andre's mistake and what is the solution to the equation?

$$-2(3x - 5) = 4(x + 3) + 8$$
$$-6x + 10 = 4x + 12 + 8$$
$$-6x + 10 = 4x + 20$$
$$10 = -2x + 20$$
$$-10 = -2x$$
$$5 = x$$

5. Choose the equation that has solutions (5, 7) and (8, 13). (Lesson 3-12)

   A. $3x - y = 8$

   B. $y = x + 2$

   C. $y - x = 5$

   D. $y = 2x - 3$

6. A length of ribbon is cut into two pieces to use in a craft project. The graph shows the length of the second piece, x, for each length of the first piece, y. (Lesson 3-9)

   a. How long is the ribbon? Explain how you know.

   b. What is the slope of the line?

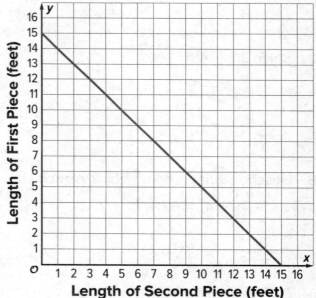

   c. Explain what the slope of the line represents and why it fits the story.

Lesson 4-5

Solving Any Linear Equation

NAME _____ DATE _____ PERIOD _____

Learning Goal Let's solve linear equations.

Warm Up
5.1 Equation Talk

Solve each equation mentally.

1. $5 - x = 8$
2. $-1 = x - 2$
3. $-3x = 9$
4. $-10 = -5x$

Activity
5.2 Trading Moves

Your teacher will give you 4 cards, each with an equation.

1. With your partner, select a card and choose who will take the first turn.

2. During your turn, decide what the next move to solve the equation should be, explain your choice to your partner, and then write it down once you both agree. Switch roles for the next move. This continues until the equation is solved.

3. Choose a second equation to solve in the same way, trading the card back and forth after each move.

4. For the last two equations, choose one each to solve and then trade with your partner when you finish to check one another's work.

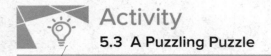

Activity

5.3 A Puzzling Puzzle

Tyler says he invented a number puzzle. He asks Clare to pick a number, and then asks her to do the following:

- Triple the number

- Subtract 7

- Double the result

- Subtract 22

- Divide by 6

Clare says she now has a -3. Tyler says her original number must have been a 3. How did Tyler know that? Explain or show your reasoning. Be prepared to share your reasoning with the class.

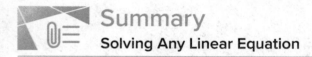

Summary

Solving Any Linear Equation

When we have an equation in one variable, there are many different ways to solve it. We generally want to make moves that get us closer to an equation like *variable = some number*. For example, $x = 5$ or $t = \frac{7}{3}$. Since there are many ways to do this, it helps to choose moves that leave fewer **terms** or factors. If we have an equation like ...

$$3t + 5 = 7,$$

adding -5 to each side will leave us with fewer terms. The equation then becomes ...

$$3t = 2.$$

Dividing each side of this equation by 3 will leave us with t by itself on the left and that ...

$$t = \frac{2}{3}.$$

Or, if we have an equation like ...

$$4(5 - a) = 12,$$

dividing each side by 4 will leave us with fewer factors on the left, ...

$$5 - a = 3.$$

NAME _____ DATE _____ PERIOD _____

Some people use the following steps to solve a linear equation in one variable:

1. Use the distributive property so that all the expressions no longer have parentheses.

2. Collect like terms on each side of the equation.

3. Add or subtract an expression so that there is a variable on just one side.

4. Add or subtract an expression so that there is just a number on the other side.

5. Multiply or divide by a number so that you have an equation that looks like *variable = some number.*

For example, suppose we want to solve $9 - 2b + 6 = -3(b + 5) + 4b$.

$$9 - 2b + 6 = -3b - 15 + 4b \quad \text{Use the distributive property.}$$
$$15 - 2b = b - 15 \quad \text{Gather like terms.}$$
$$15 = 3b - 15 \quad \text{Add } 2b \text{ to each side.}$$
$$30 = 3b \quad \text{Add 15 to each side.}$$
$$10 = b \quad \text{Divide each side by 3.}$$

Following these steps will always work, although it may not be the most efficient method. From lots of experience, we learn when to use different approaches.

Glossary

term

Practice
Solving Any Linear Equation

1. Solve each of these equations. Explain or show your reasoning.

   $2(x + 5) = 3x + 1$ $\qquad\qquad$ $3y - 4 = 6 - 2y$ $\qquad\qquad$ $3(n + 2) = 9(6 - n)$

2. Clare was solving an equation, but when she checked her answer she saw her solution was incorrect. She knows she made a mistake, but she can't find it. Where is Clare's mistake and what is the solution to the equation?

$$12(5 + 2y) = 4y - (5 - 9y)$$
$$72 + 24y = 4y - 5 - 9y$$
$$72 + 24y = -5y - 5$$
$$24y = -5y - 77$$
$$29y = -77$$
$$y = \frac{-77}{29}$$

NAME _____ DATE _____ PERIOD _____

3. Solve each equation, and check your solution.

$\frac{1}{9}(2m - 16) = \frac{1}{3}(2m + 4)$

-4(r + 2) = 4(2 − 2r)

12(5 + 2y) = 4y − (6 − 9y)

4. Here is the graph of a linear equation. (Lesson 3-13)

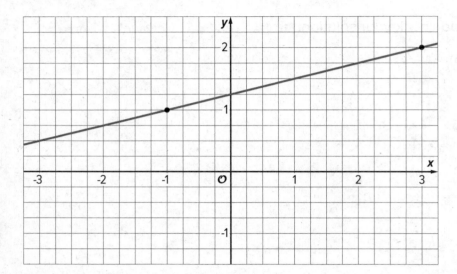

Select **all** true statements about the line and its equation.

A. One solution of the equation is (3, 2).

B. One solution of the equation is (-1, 1).

C. One solution of the equation is $\left(1, \dfrac{3}{2}\right)$.

D. There are 2 solutions.

E. There are infinitely many solutions.

F. The equation of the line is $y = \dfrac{1}{4}x + \dfrac{5}{4}$.

G. The equation of the line is $y = \dfrac{5}{4}x + \dfrac{1}{4}$.

5. A participant in a 21-mile walkathon walks at a steady rate of 3 miles per hour. He thinks, "The relationship between the number of miles left to walk and the number of hours I already walked can be represented by a line with slope -3." Do you agree with his claim? Explain your reasoning. (Lesson 3-9)

Lesson 4-6

Strategic Solving

NAME _____ DATE _____ PERIOD _____

Learning Goal Let's solve linear equations like a boss.

Warm Up
6.1 Equal Perimeters

The triangle and the square have equal perimeters.

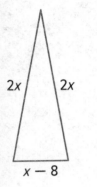

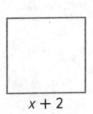

1. Find the value of x.

2. What is the perimeter of each of the figures?

Activity
6.2 Predicting Solutions

Without solving, identify whether these equations have a solution that is positive, negative, or zero.

1. $\frac{x}{6} = \frac{3x}{4}$

2. $7x = 3.25$

3. $7x = 32.5$

4. $3x + 11 = 11$

5. $9 - 4x = 4$

6. $-8 + 5x = -20$

7. $-\frac{1}{2}(-8 + 5x) = -20$

NAME _____ DATE _____ PERIOD _____

Activity

6.3 Which Would You Rather Solve?

Here are a lot of equations:

A. $-\frac{5}{6}(8 + 5b) = 75 + \frac{5}{3}b$

F. $3(c - 1) + 2(3c + 1) = -(3c + 1)$

B. $-\frac{1}{2}(t + 3) - 10 = -6.5$

G. $\frac{4m - 3}{4} = -\frac{9 + 4m}{8}$

C. $\frac{10 - v}{4} = 2(v + 17)$

H. $p - 5(p + 4) = p - (8 - p)$

D. $2(4k + 3) - 13 = 2(18 - k) - 13$

I. $2(2q + 1.5) = 18 - q$

E. $\frac{n}{7} - 12 = 5n + 5$

J. $2r + 49 = -8(-r - 5)$

1. Without solving, identify 3 equations that you think would be least difficult to solve and 3 equations you think would be most difficult to solve. Be prepared to explain your reasoning.

2. Choose 3 equations to solve. At least one should be from your "least difficult" list and one should be from your "most difficult" list.

Are you ready for more?

Mai gave half of her brownies, and then half a brownie more, to Kiran. Then she gave half of what was left, and half a brownie more, to Tyler. That left her with one remaining brownie. How many brownies did she have to start with?

Summary
Strategic Solving

Sometimes we are asked to solve equations with a lot of things going on on each side. For example, . . .

$$x - 2(x + 5) = \frac{3(2x - 20)}{6}$$

This equation has variables on each side, parentheses, and even a fraction to think about. Before we start distributing, let's take a closer look at the fraction on the right side. The expression $2x - 20$ is being multiplied by 3 and divided by 6, which is the same as just dividing by 2, so we can re-write the equation as . . .

$$x - 2(x + 5) = \frac{2x - 20}{2}$$

But now it's easier to see that all the terms on the numerator of right side are divisible by 2, which means we can re-write the right side again as . . .

$$x - 2(x + 5) = x - 10$$

At this point, we could do some distribution and then collect like terms on each side of the equation. Another choice would be to use the structure of the equation. Both the left and the right side have something being subtracted from x. But, if the two sides are equal, that means the "something" being subtracted on each side must also be equal. Thinking this way, the equation can now be re-written with less terms as . . .

$$2(x + 5) = 10$$

Only a few steps left! But what can we tell about the solution to this problem right now? Is it positive? Negative? Zero?

Well, the 2 and the 5 multiplied together are 10, so that means the 2 and the x multiplied together cannot have a positive or a negative value. Finishing the steps we have . . .

$$
\begin{aligned}
2(x + 5) &= 10 \\
x + 5 &= 5 \quad \text{Divide each side by 2.} \\
x &= 0 \quad \text{Subtract 5 from each side.}
\end{aligned}
$$

Neither positive nor negative. Just as predicted.

NAME _____ DATE _____ PERIOD _____

Practice
Strategic Solving

1. Solve each of these equations. Explain or show your reasoning.

   a. $2b + 8 - 5b + 3 = \text{-}13 + 8b - 5$

   b. $2x + 7 - 5x + 8 = 3(5 + 6x) - 12x$

   c. $2c - 3 = 2(6 - c) + 7c$

2. Solve each equation and check your solution.

   a. $\text{-}3w - 4 = w + 3$

   b. $3(3 - 3x) = 2(x + 3) - 30$

   c. $\frac{1}{3}(z + 4) - 6 = \frac{2}{3}(5 - z)$

3. Elena said the equation $9x + 15 = 3x + 15$ has no solutions because $9x$ is greater than $3x$. Do you agree with Elena? Explain your reasoning.

4. The table gives some sample data for two quantities, x and y, that are in a proportional relationship. (Lesson 3-3)

x	y
14	21
64	
	39
1	

a. Complete the table.

b. Write an equation that represents the relationship between x and y shown in the table.

c. Graph the relationship. Use a scale for the axes that shows all the points in the table.

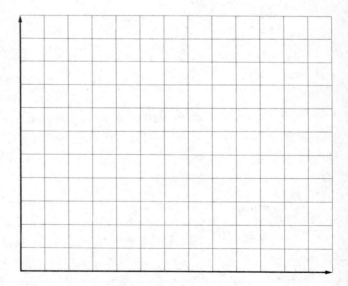

Lesson 4-7

All, Some, or No Solutions

NAME _____ DATE _____ PERIOD _____

Learning Goal Let's think about how many solutions an equation can have.

 ## Warm Up
7.1 Which One Doesn't Belong: Equations

Which one doesn't belong?

1. $5 + 7 = 7 + 5$

2. $5 \cdot 7 = 7 \cdot 5$

3. $2 = 7 - 5$

4. $5 - 7 = 7 - 5$

Activity

7.2 Thinking About Solutions

1. Sort these equations into the two types: true for all values and true for no values.

$n = n$

$2t + 6 = 2(t + 3)$

$3(n + 1) = 3n + 1$

$\frac{1}{4}(20d + 4) = 5d$

$5 - 9 + 3x = \text{-}10 + 6 + 3x$

$\frac{1}{2} + x = \frac{1}{3} + x$

$y \cdot \text{-}6 \cdot \text{-}3 = 2 \cdot y \cdot 9$

$v + 2 = v - 2$

2. Write the other side of this equation so that this equation is true for all values of u.

$6(u - 2) + 2 =$

3. Write the other side of this equation so that this equation is true for no values of u.

$6(u - 2) + 2 =$

NAME _____ DATE _____ PERIOD _____

Are you ready for more?

Consecutive numbers follow one right after the other. An example of three consecutive numbers is 17, 18, and 19. Another example is -100, -99, -98.

How many sets of two or more consecutive positive integers can be added to obtain a sum of 100?

Activity

7.3 What's the Equation?

1. Complete each equation so that it is true for all values of x.

   a. $3x + 6 = 3(x + $ _____ $)$

   b. $x - 2 = $ -$($ _____ $- x)$

   c. $\dfrac{15x - 10}{5} = $ _____ $- 2$

2. Complete each equation so that it is true for no values of x.

   a. $3x + 6 = 3(x + $ _____ $)$

   b. $x - 2 = $ -$($ _____ $- x)$

   c. $\dfrac{15x - 10}{5} = $ _____ $- 2$

3. Describe how you know whether an equation will be true for all values of x or true for no values of x.

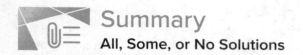

Summary

All, Some, or No Solutions

An equation is a statement that two expressions have an equal value. The equation $2x = 6$ . . .

- is a true statement if x is 3:
  $2 \cdot 3 = 6$

- is a false statement if x is 4:
  $2 \cdot 4 = 6$

The equation $2x = 6$ has *one and only one solution*, because there is only one number (3) that you can double to get 6.

Some equations are true no matter what the value of the variable is. For example, $2x = x + x$ . . .
is always true, because if you double a number, that will always be the same as adding the number to itself.
Equations like $2x = x + x$ have an *infinite number of solutions*. We say it is true for all values of x.

Some equations have *no solutions*.
For example, $x = x + 1$ . . .
has no solutions, because no matter what the value of x is, it can't equal one more than itself.

When we solve an equation, we are looking for the values of the variable that make the equation true. When we try to solve the equation, we make allowable moves assuming it *has* a solution.
Sometimes we make allowable moves and get an equation like this: $8 = 7$. This statement is false, so it must be that the original equation had no solution at all.

NAME _____ DATE _____ PERIOD _____

Practice
All, Some, or No Solutions

1. For each equation, decide if it is always true or never true.

   a. $x - 13 = x + 1$

   b. $x + \dfrac{1}{2} = x - \dfrac{1}{2}$

   c. $2(x + 3) = 5x + 6 - 3x$

   d. $x - 3 = 2x - 3 - x$

   e. $3(x - 5) = 2(x - 5) + x$

2. Mai says that the equation $2x + 2 = x + 1$ has no solution because the left hand side is double the right hand side. Do you agree with Mai? Explain your reasoning.

3. a. Write the other side of this equation so it's true for all values of x:
   $$\tfrac{1}{2}(6x - 10) - x =$$

   b. Write the other side of this equation so it's true for no values of x:
   $$\tfrac{1}{2}(6x - 10) - x =$$

4. Here is an equation that is true for all values of x: $5(x + 2) = 5x + 10$. Elena saw this equation and says she can tell $20(x + 2) + 31 = 4(5x + 10) + 31$ is also true for any value of x. How can she tell? Explain your reasoning.

5. Elena and Lin are trying to solve $\frac{1}{2}x + 3 = \frac{7}{2}x + 5$. Describe the change they each make to each side of the equation. (Lesson 4-4)

   a. Elena's first step is to write $3 = \frac{7}{2}x - \frac{1}{2}x + 5$.

   b. Lin's first step is to write $x + 6 = 7x + 10$.

6. Solve each equation and check your solution. (Lesson 4-6)

   $$3x - 6 = 4(2 - 3x) - 8x \qquad \frac{1}{2}z + 6 = \frac{3}{2}(z + 6) \qquad 9 - 7w = 8w + 8$$

7. The point (-3, 6) is on a line with a slope of 4. (Lesson 3-12)

   a. Find two more points on the line.

   b. Write an equation for the line.

Lesson 4-8

How Many Solutions?

NAME _____ DATE _____ PERIOD _____

Learning Goal Let's solve equations with different numbers of solutions.

Warm Up
8.1 Matching Solutions

Consider the unfinished equation $12(x - 3) + 18 = $ _____.
Match the following expressions with the number of solutions the equation
would have with that expression on the right hand side.

1. $6(2x - 3)$ • One solution

2. $4(3x - 3)$ • No solutions

3. $4(2x - 3)$ • All solutions

Activity
8.2 Thinking About Solutions Some More

Your teacher will give you some cards.

1. With your partner, solve each equation.

2. Then, sort them into categories.

3. Describe the defining characteristics of those categories and be prepared
   to share your reasoning with the class.

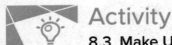

Activity

8.3 Make Use of Structure

For each equation, determine whether it has no solutions, exactly one solution, or is true for all values of x (and has infinitely many solutions). If an equation has one solution, solve to find the value of x that makes the statement true.

1. a. $6x + 8 = 7x + 13$

   b. $6x + 8 = 2(3x + 4)$

   c. $6x + 8 = 6x + 13$

2. a. $\frac{1}{4}(12 - 4x) = 3 - x$

   b. $x - 3 = 3 - x$

   c. $x - 3 = 3 + x$

3. a. $-5x - 3x + 2 = -8x + 2$

   b. $-5x - 3x - 4 = -8x + 2$

   c. $-5x - 4x - 2 = -8x + 2$

4. a. $4(2x - 2) + 2 = 4(x - 2)$

   b. $4x + 2(2x - 3) = 8(x - 1)$

   c. $4x + 2(2x - 3) = 4(2x - 2) + 2$

NAME _____ DATE _____ PERIOD _____

5. a. $x - 3(2 - 3x) = 2(5x + 3)$

   b. $x - 3(2 + 3x) = 2(5x - 3)$

   c. $x - 3(2 - 3x) = 2(5x - 3)$

6. What do you notice about equations with one solution? How is this different from equations with no solutions and equations that are true for every x?

Are you ready for more?

Consecutive numbers follow one right after the other. An example of three consecutive numbers is 17, 18, and 19. Another example is -100, -99, -98.

1. Choose any set of three consecutive numbers. Find their average. What do you notice?

2. Find the average of another set of three consecutive numbers. What do you notice?

3. Explain why the thing you noticed must always work, or find a counterexample.

Summary
How Many Solutions?

Sometimes it's possible to look at the structure of an equation and tell if it has infinitely many solutions or no solutions. For example, look at . . .

$$2(12x + 18) + 6 = 18x + 6(x + 7).$$

Using the distributive property on the left and right sides, we get . . .

$$24x + 36 + 6 = 18x + 6x + 42.$$

From here, collecting like terms gives us . . .

$$24x + 42 = 24x + 42.$$

Since the left and right sides of the equation are the same, we know that this equation is true for any value of x without doing any more moves!

Similarly, we can sometimes use structure to tell if an equation has no solutions. For example, look at . . .

$$6(6x + 5) = 12(3x + 2) + 12.$$

If we think about each move as we go, we can stop when we realize there is no solution:

$\frac{1}{6} \cdot 6(6x + 5) = \frac{1}{6} \cdot (12(3x + 2) + 12)$	Multiply each side by $\frac{1}{6}$.
$6x + 5 = 2(3x + 2) + 2$	Distribute $\frac{1}{6}$ on the right side.
$6x + 5 = 6x + 4 + 2$	Distribute 2 on the right side.

The last move makes it clear that the **constant terms** on each side, 5 and $4 + 2$, are not the same. Since adding 5 to an amount is always less than adding $4 + 2$ to that same amount, we know there are no solutions.

Doing moves to keep an equation balanced is a powerful part of solving equations, but thinking about what the structure of an equation tells us about the solutions is just as important.

Glossary

coefficient

constant term

NAME _____ DATE _____ PERIOD _____

Practice
How Many Solutions?

1. Lin was looking at the equation $2x - 32 + 4(3x - 2{,}462) = 14x$. She said, "I can tell right away there are no solutions, because on the left side, you will have $2x + 12x$ and a bunch of constants, but you have just $14x$ on the right side." Do you agree with Lin? Explain your reasoning.

2. Han was looking at the equation $6x - 4 + 2(5x + 2) = 16x$. He said, "I can tell right away there are no solutions, because on the left side, you will have $6x + 10x$ and a bunch of constants, but you have just $16x$ on the right side." Do you agree with Han? Explain your reasoning.

3. Decide whether each equation is true for all, one, or no values of x.

   a. $6x - 4 = -4 + 6x$

   b. $4x - 6 = 4x + 3$

   c. $-2x + 4 = -3x + 4$

4. Solve each of these equations. Explain or show your reasoning. **(Lesson 4-4)**

   a. $3(x - 5) = 6$

   b. $2\left(x - \dfrac{2}{3}\right) = 0$

   c. $4x - 5 = 2 - x$

5. The points (-2, 0) and (0, -6) are each on the graph of a linear equation. Is (2, 6) also on the graph of this linear equation? Explain your reasoning. (Lesson 3-13)

6. In the picture triangle $A'B'C'$ is an image of triangle ABC after a rotation. The center of rotation is E. (Lesson 1-7)

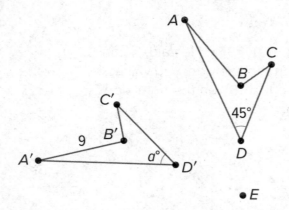

a. What is the length of side AB? Explain how you know.

b. What is the measure of angle D'? Explain how you know.

Lesson 4-9

When Are They the Same?

NAME _____ DATE _____ PERIOD _____

Learning Goal Let's use equations to think about situations.

 ## Warm Up
9.1 Which Would You Choose?

If you were babysitting, would you rather

- charge $5 for the first hour and $8 for each additional hour?

Or

- charge $15 for the first hour and $6 for each additional hour?

Explain your reasoning.

Activity
9.2 Water Tanks

The amount of water in two tanks every 5 minutes is shown in the table.

Time (minutes)	Tank 1 (liters)	Tank 2 (liters)
0	25	1000
5	175	900
10	325	800
15	475	700
20	625	600
25	775	500
30	925	400
35	1075	300
40	1225	200
45	1375	100
50	1525	0

1. Describe what is happening in each tank. Either draw a picture, say it verbally, or write a few sentences.

2. Use the table to estimate when the tanks will have the same amount of water.

3. The amount of water (in liters) in tank 1 after t minutes is $30t + 25$. The amount of water (in liters) in tank 2 after t minutes is $-20t + 1000$. Find the time when the amount of water will be equal.

NAME _____ DATE _____ PERIOD _____

Activity

9.3 Elevators

A building has two elevators that both go above and below ground.

At a certain time of day, the travel time it takes elevator A to reach height h in meters is $0.8h + 16$ seconds.

The travel time it takes elevator B to reach height h in meters is $-0.8h + 12$ seconds.

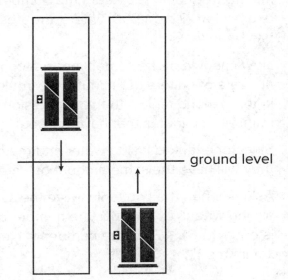

ground level

1. What is the height of each elevator at this time?

2. How long would it take each elevator to reach ground level at this time?

3. If the two elevators travel toward one another, at what height do they pass each other? How long would it take?

4. If you are on an underground parking level 14 meters below ground, which elevator would reach you first?

Are you ready for more?

1. In a two-digit number, the ones digit is twice the tens digit. If the digits are reversed, the new number is 36 more than the original number. Find the number.

2. The sum of the digits of a two-digit number is 11. If the digits are reversed, the new number is 45 less than the original number. Find the number.

3. The sum of the digits in a two-digit number is 8. The value of the number is 4 less than 5 times the ones digit. Find the number.

Imagine a full 1,500 liter water tank that springs a leak, losing 2 liters per minute. We could represent the number of liters left in the tank with the expression $-2x + 1,500$, where x represents the number of minutes the tank has been leaking.

Now imagine at the same time, a second tank has 300 liters and is being filled at a rate of 6 liters per minute. We could represent the amount of water in liters in this second tank with the expression $6x + 300$, where x represents the number of minutes that have passed.

Since one tank is losing water and the other is gaining water, at some point they will have the same amount of water—but when?

Asking when the two tanks have the same number of liters is the same as asking when $-2x + 1,500$ (the number of liters in the first tank after x minutes) is equal to $6x + 300$ (the number of liters in the second tank after x minutes), . . .

$-2x + 1,500 = 6x + 300$.

Solving for x gives us $x = 150$ minutes.
So after 150 minutes, the number of liters of the first tank is equal to the number of liters of the second tank.

But how much water is actually in each tank at that time? Since both tanks have the same number of liters after 150 minutes, we could substitute $x = 150$ minutes into either expression.

Using the expression for the first tank, we get $-2(150) + 1,500$ which is equal to $-300 + 1,500$, or 1,200 liters.

If we use the expression for the second tank, we get $6(150) + 300$, or just $900 + 300$, which is also 1,200 liters.
That means that after 150 minutes, each tank has 1,200 liters.

NAME _____ DATE _____ PERIOD _____

Practice
When Are They the Same?

1. Cell phone Plan A costs $70 per month and comes with a free $500 phone. Cell phone Plan B costs $50 per month but does not come with a phone. If you buy the $500 phone and choose Plan B, how many months is it until your cost is the same as Plan A's?

2. Priya and Han are biking in the same direction on the same path.

   a. Han is riding at a constant speed of 16 miles per hour. Write an expression that shows how many miles Han has gone after t hours.

   b. Priya started riding a half hour before Han. If Han has been riding for t hours, how long has Priya been riding?

   c. Priya is riding at a constant speed of 12 miles per hour. Write an expression that shows how many miles Priya has gone after Han has been riding for t hours.

   d. Use your expressions to find when Han and Priya meet.

3. Which story matches the equation $-6 + 3x = 2 + 4x$?

   (A.) At 5 p.m., the temperatures recorded at two weather stations in Antarctica are -6 degrees and 2 degrees. The temperature changes at the same constant rate, x degrees per hour, throughout the night at both locations. The temperature at the first station 3 hours after this recording is the same as the temperature at the second station 4 hours after this recording.

   (B.) Elena and Kiran play a card game. Every time they collect a pair of matching cards, they earn x points. At one point in the game, Kiran has -6 points and Elena has 2 points. After Elena collects 3 pairs and Kiran collects 4 pairs, they have the same number of points.

4. For what value of x do the expressions $\frac{2}{3}x + 2$ and $\frac{4}{3}x - 6$ have the same value?

5. Decide whether each equation is true for all, one, or no values of x. (Lesson 4-8)

   a. $2x + 8 = \text{-}3.5x + 19$

   b. $9(x - 2) = 7x + 5$

   c. $3(3x + 2) - 2x = 7x + 6$

6. Solve each equation. Explain your reasoning. (Lesson 4-6)

   a. $3d + 16 = \text{-}2(5 - 3d)$

   b. $2k - 3(4 - k) = 3k + 4$

   c. $\dfrac{3y - 6}{9} = \dfrac{4 - 2y}{\text{-}3}$

7. Describe a rigid transformation that takes Polygon A to Polygon B. (Lesson 1-7)

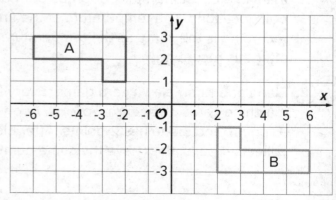

Lesson 4-10

On or Off the Line?

NAME _____ DATE _____ PERIOD _____

Learning Goal Let's interpret the meaning of points in a coordinate plane.

Warm Up

10.1 Which One Doesn't Belong: Lines in the Plane

Which one doesn't belong? Explain your reasoning.

A

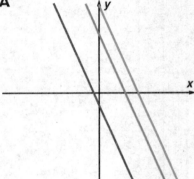

B

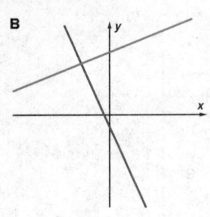

C

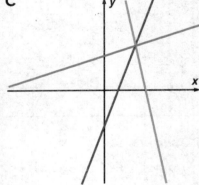

D

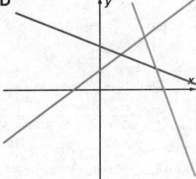

10.2 Pocket Full of Change

Jada told Noah that she has $2 worth of quarters and dimes in her pocket and 17 coins all together. She asked him to guess how many of each type of coin she has.

1. Here is a table that shows some combinations of quarters and dimes that are worth $2. Complete the table.

Number of Quarters	Number of Dimes
0	20
4	
	0
	5

2. Here is a graph of the relationship between the number of quarters and the number of dimes when there are a total of 17 coins.

   a. What does Point A represent?

   b. How much money, in dollars, is the combination represented by Point A worth?

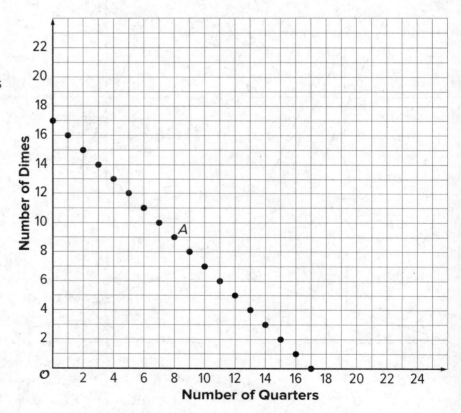

3. Is it possible for Jada to have 4 quarters and 13 dimes in her pocket? Explain how you know.

4. How many quarters and dimes must Jada have? Explain your reasoning.

NAME _____ DATE _____ PERIOD _____

Activity
10.3 Making Signs

Clare and Andre are making signs for all the lockers as part of the decorations for the upcoming spirit week. Yesterday, Andre made 15 signs and Clare made 5 signs. Today, they need to make more signs. Each person's progress today is shown in the coordinate plane.

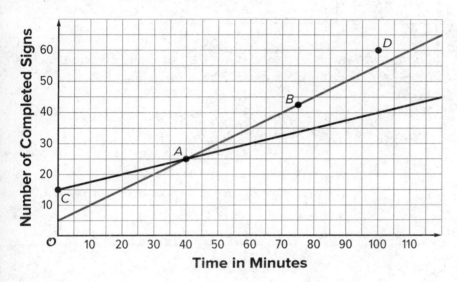

Based on the lines, mark the statements as true or false for each person.

Point	What It Says	Clare	Andre
A	At 40 minutes, I have 25 signs completed.		
B	At 75 minutes, I have 42 and a half signs completed.		
C	At 0 minutes, I have 15 signs completed.		
D	At 100 minutes, I have 60 signs completed.		

- 4 toothpicks make 1 square.
- 7 toothpicks make 2 squares.
- 10 toothpicks make 3 squares.

Do you see a pattern? If so, how many toothpicks would you need to make 10 squares according to your pattern? Can you represent your pattern with an expression?

Summary

On or Off the Line?

We studied linear relationships in an earlier unit. We learned that values of x and y that make an equation true correspond to points (x, y) on the graph. For example, if we have x pounds of flour that costs $0.80 per pound and y pounds of sugar that costs $0.50 per pound, and the total cost is $9.00, then we can write an equation like this to represent the relationship between x and y:

$$0.8x + 0.5y = 9$$

Since 5 pounds of flour costs $4.00 and 10 pounds of sugar costs $5.00, we know that $x = 5$, $y = 10$ is a solution to the equation, and the point $(5, 10)$ is a point on the graph. The line shown is the graph of the equation:

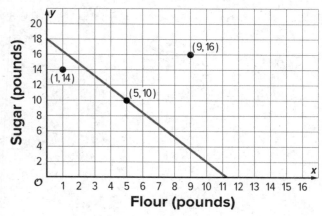

NAME _____ DATE _____ PERIOD _____

Notice that there are two points shown that are not on the line. What do they mean in the context? The point (1, 14) means that there is 1 pound of flour and 14 pounds of sugar. The total cost for this is $0.8 \cdot 1 + 0.5 \cdot 14$ or $7.80.
Since the cost is not $9.00, this point is not on the graph. Likewise, 9 pounds of flour and 16 pounds of sugar costs $0.8 \cdot 9 + 0.5 \cdot 16$ or $15.20, so the other point is not on the graph either.

Suppose we also know that the flour and sugar together weigh 15 pounds. That means that . . .

$$x + y = 15.$$

If we draw the graph of this equation on the same coordinate plane, we see it passes through two of the three labeled points:

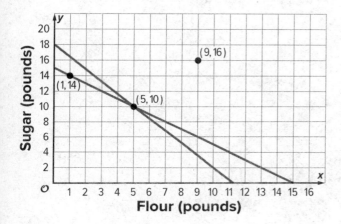

The point (1, 14) is on the graph of $x + y = 15$ because $1 + 14 = 15$. Similarly, $5 + 10 = 15$. But $9 + 16 \neq 15$, so (9, 16) is *not* on the graph of $x + y = 15$.

In general, if we have two lines in the coordinate plane,

• The coordinates of a point that is on both lines makes both equations true.

• The coordinates of a point on only one line makes only one equation true.

• The coordinates of a point on neither line make both equations false.

Practice
On or Off the Line?

1. a. Match the lines m and n to the statements they represent:

- A set of points where the coordinates of each point have a sum of 2

- A set of points where the y-coordinate of each point is 10 less than its x-coordinate

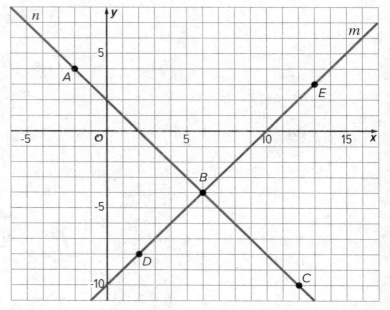

b. Match the labeled points on the graph to statements about their coordinates.

- Two numbers with a sum of 2

- Two numbers where the y-coordinate is 10 less than the x-coordinate

- Two numbers with a sum of 2 and where the y-coordinate is 10 less than the x-coordinate

2. Here is an equation: 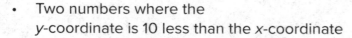 $4x - 4 = 4x + \_\_$. What could you write in the blank so the equation would be true for: (Lesson 4-7)

a. No values of x

b. All values of x

c. One value of x

NAME _____ DATE _____ PERIOD _____

3. Mai earns $7 per hour mowing her neighbors' lawns. She also earned $14 for hauling away bags of recyclables for some neighbors.

Priya babysits her neighbor's children. The table shows the amount of money m she earns in h hours. Priya and Mai have agreed to go to the movies the weekend after they have earned the *same* amount of money for the *same* number of work hours.

h	m
1	$8.40
2	$16.80
4	$33.60

a. How many hours do they each have to work before they go to the movies?

b. How much will each of them have earned?

c. Explain where the solution can be seen in tables of values, graphs, and equations that represent Priya's and Mai's hourly earnings.

4. For each equation, explain what you could do first to each side of the equation so that there would be no fractions. You do not have to solve the equations (unless you want more practice). **(Lesson 4-6)**

a. $\dfrac{3x - 4}{8} = \dfrac{x + 2}{3}$

b. $\dfrac{3(2 - r)}{4} = \dfrac{3 + r}{6}$

c. $\dfrac{4p + 3}{8} = \dfrac{p + 2}{4}$

d. $\dfrac{2(a - 7)}{15} = \dfrac{a + 4}{6}$

5. The owner of a new restaurant is ordering tables and chairs. He wants to have only tables for 2 and tables for 4. The total number of people that can be seated in the restaurant is 120.

  a. Describe some possible combinations of 2-seat tables and 4-seat tables that will seat 120 customers. Explain how you found them.

  b. Write an equation to represent the situation. What do the variables represent?

  c. Create a graph to represent the situation.

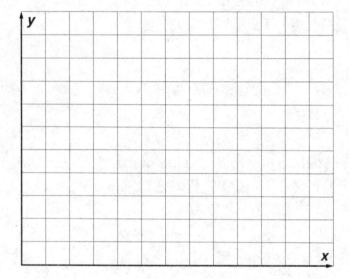

  d. What does the slope tell us about the situation?

  e. Interpret the *x*- and *y*- intercepts in the situation.

Lesson 4-11

On Both of the Lines

NAME _____ DATE _____ PERIOD _____

Learning Goal Let's use lines to think about situations.

Warm Up
11.1 Notice and Wonder: Bugs Passing in the Night

What do you notice? What do you wonder?

Activity

11.2 Bugs Passing in the Night, Continued

A different ant and ladybug are a certain distance apart, and they start walking toward each other. The graph shows the ladybug's distance from its starting point over time and the labeled point (2.5, 10) indicates when the ant and the ladybug pass each other.

The ant is walking 2 centimeters per second.

1. Write an equation representing the relationship between the ant's distance from the ladybug's starting point and the amount of time that has passed.

2. If you haven't already, draw the graph of your equation on the same coordinate plane.

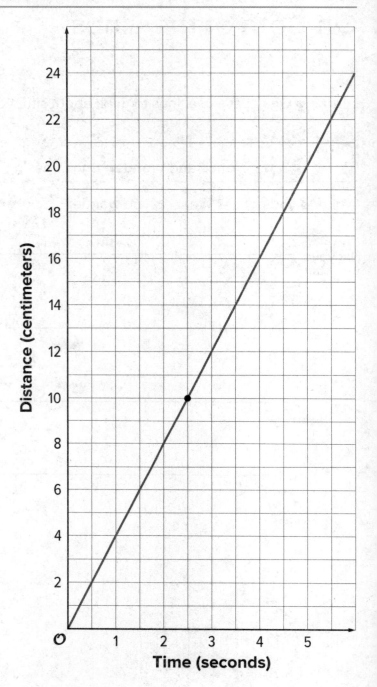

NAME _____ DATE _____ PERIOD _____

Activity
11.3 A Close Race

Elena and Jada were racing 100 meters on their bikes. Both racers started at the same time and rode at a constant speed. Here is a table that gives information about Jada's bike race.

Time from Start (seconds)	Distance from Start (meters)
6	36
9	54

1. Graph the relationship between distance and time for Jada's bike race. Make sure to label and scale the axes appropriately.

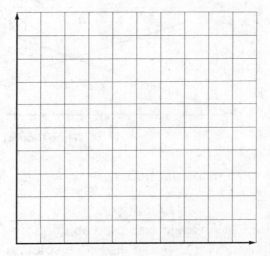

2. Elena traveled the entire race at a steady 6 meters per second. On the same set of axes, graph the relationship between distance and time for Elena's bike race.

3. Who won the race?

Summary
On Both of the Lines

The solutions to an equation correspond to points on its graph. For example, if Car A is traveling 75 miles per hour and passes a rest area when $t = 0$, then the distance in miles it has traveled from the rest area after t hours is $d = 75t$.

The point (2, 150) is on the graph of this equation because $150 = 75 \cdot 2$: two hours after passing the rest area, the car has traveled 150 miles.

If you have *two* equations, you can ask whether there is an ordered pair that is a solution to *both* equations simultaneously. For example, if Car B is traveling towards the rest area and its distance from the rest area is $d = 14 - 65t$, we can ask if there is ever a time when the distance of Car A from the rest area is the same as the distance of Car B from the rest area. If the answer is "yes", then the solution will correspond to a point that is on both lines.

Looking at the coordinates of the intersection point, we see that Car A and Car B will both be 7.5 miles from the rest area after 0.1 hours (which is 6 minutes).

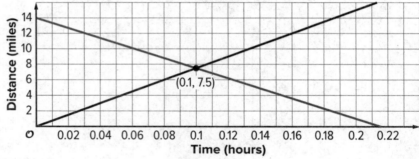

Now suppose another car, Car C, had also passed the rest stop at time $t = 0$ and traveled in the same direction as Car A, also going 75 miles per hour.

- It's equation would also be $d = 75t$.

- Any solution to the equation for Car A would also be a solution for Car C, and any solution to the equation for Car C would also be a solution for Car A.

- The line for Car C would land right on top of the line for Car A. In this case, every point on the graphed line is a solution to both equations, so that there are infinitely many solutions to the question "when are Car A and Car C the same distance from the rest stop?" This would mean that Car A and Car C were side by side for their whole journey.

When we have two linear equations that are equivalent to each other, like $y = 3x + 2$ and $2y = 6x + 4$, we will get two lines that are "right on top" of each other. Any solution to one equation is also solution to the other, so these two lines intersect at infinitely many points.

NAME _____ DATE _____ PERIOD _____

Practice
On Both of the Lines

1. Diego has $11 and begins saving $5 each week toward buying a new phone. At the same time that Diego begins saving, Lin has $60 and begins spending $2 per week on supplies for her art class. Is there a week when they have the same amount of money? How much do they have at that time?

2. Use a graph to find x and y values that make both $y = \frac{-2}{3}x + 3$ and $y = 2x - 5$ true.

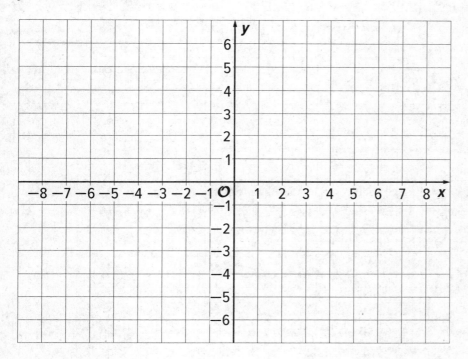

3. The point where the graphs of two equations intersect has y-coordinate 2. One equation is $y = -3x + 5$. Find the other equation if its graph has a slope of 1.

4. A farm has chickens and cows. All the cows have 4 legs and all the chickens have 2 legs. All together, there are 82 cow and chicken legs on the farm. Complete the table to show some possible combinations of chickens and cows to get 82 total legs.

Number of Chickens (x)	Number of Cows (y)
35	
7	
	10
19	
	5

Here is a graph that shows possible combinations of chickens and cows that add up to 30 animals. If the farm has 30 chickens and cows, and there are 82 chicken and cow legs all together, then how many chickens and how many cows could the farm have? (Lesson 4-10)

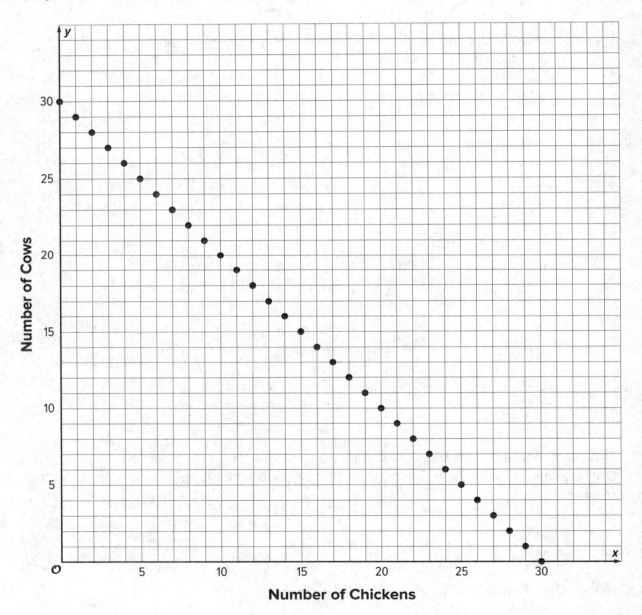

Number of Cows

Number of Chickens

Lesson 4-12

Systems of Equations

NAME _____ DATE _____ PERIOD _____

Learning Goal Let's learn what a system of equations is.

 **Warm Up**
12.1 Milkshakes

Diego and Lin are drinking milkshakes. Lin starts with 12 ounces and drinks $\frac{1}{4}$ ounce per second. Diego starts with 20 ounces and drinks $\frac{2}{3}$ ounce per second.

1. How long will it take Lin and Diego to finish their milkshakes?

2. Without graphing, explain what the graphs in this situation would look like. Think about slope, intercepts, axis labels, units, and intersection points to guide your thinking.

3. Discuss your description with your partner. If you disagree, work to reach an agreement.

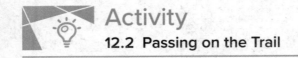

Activity

12.2 Passing on the Trail

There is a hiking trail near the town where Han and Jada live that starts at a parking lot and ends at a lake. Han and Jada both decide to hike from the parking lot to the lake and back, but they start their hikes at different times.

At the time that Han reaches the lake and starts to turn back, Jada is 0.6 miles away from the parking lot and hiking at a constant speed of 3.2 miles per hour towards the lake. Han's distance, d, from the parking lot can be expressed as $d = -2.4t + 4.8$, where t represents the time in hours since he left the lake.

1. What is an equation for Jada's distance from the parking lot as she heads toward the lake?

2. Draw both graphs: one representing Han's equation and one representing Jada's equation. It is important to be very precise! Be careful, work in pencil, and use a ruler.

3. Find the point where the two graphs intersect each other. What are the coordinates of this point?

4. What do the coordinates mean in this situation?

5. What has to be true about the relationship between these coordinates and Jada's equation?

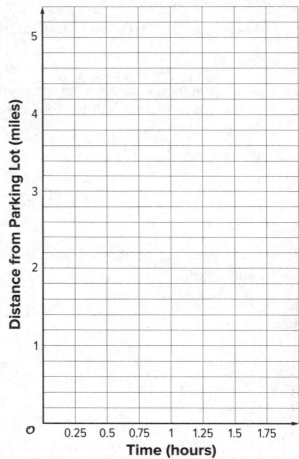

6. What has to be true about the relationship between these coordinates and Han's equation?

NAME _____ DATE _____ PERIOD _____

Activity

12.3 Stacks of Cups

A stack of n small cups has a height, h, in centimeters of $h = 1.5n + 6$.
A stack of n large cups has a height, h, in centimeters of $h = 1.5n + 9$.

1. Graph the equations for each cup on the same set of axes. Make sure to label the axes and decide on an appropriate scale.

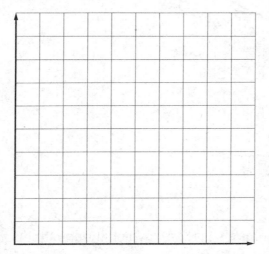

2. For what number of cups will the two stacks have the same height?

Summary

Systems of Equations

A **system of equations** is a set of 2 (or more) equations where the variables represent the same unknown values. For example, suppose that two different kinds of bamboo are planted at the same time. Plant A starts at 6 ft tall and grows at a constant rate of $\frac{1}{4}$ foot each day. Plant B starts at 3 ft tall and grows at a constant rate of $\frac{1}{2}$ foot each day. We can write equations $y = \frac{1}{4}x + 6$ for Plant A and $y = \frac{1}{2}x + 3$ for Plant B, where x represents the number of days after being planted, and y represents height. We can write this system of equations.

$$\begin{cases} y = \frac{1}{4}x + 6 \\ y = \frac{1}{2}x + 3 \end{cases}$$

Solving a system of equations means to find the values of x and y that make both equations true at the same time. One way we have seen to find the solution to a system of equations is to graph both lines and find the intersection point. The intersection point represents the pair of x and y values that make both equations true.

Here is a graph for the bamboo example:

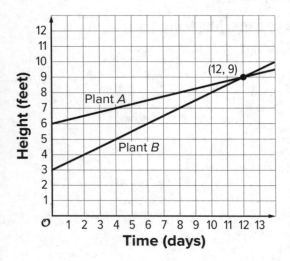

The solution to this system of equations is (12, 9), which means that both bamboo plants will be 9 feet tall after 12 days.

We have seen systems of equations that have no solutions, one solution, and infinitely many solutions.

*   When the lines do not intersect, there is no solution. (Lines that do not intersect are *parallel*.)

*   When the lines intersect once, there is one solution.

*   When the lines are right on top of each other, there are infinitely many solutions.

In future lessons, we will see that some systems cannot be easily solved by graphing, but can be easily solved using algebra.

Glossary

system of equations

NAME _____ DATE _____ PERIOD _____

 Practice
Systems of Equations

1. Here is the graph for one equation in a system of equations:

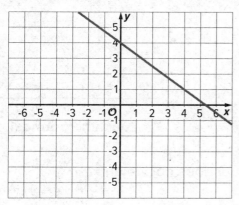

a. Write a second equation for the system so it has infinitely many solutions.

b. Write a second equation whose graph goes through (0, 1) so the system has no solutions.

c. Write a second equation whose graph goes through (0, 2) so the system has one solution at (4, 1).

2. Create a second equation so the system has no solutions.

$$\left\{ y = \frac{3}{4}x - 4 \right.$$

3. Andre is in charge of cooking broccoli and zucchini for a large group. He has to spend all $17 he has and carry 10 pounds of veggies. Zucchini costs $1.50 per pound and broccoli costs $2 per pound. One graph shows combinations of zucchini and broccoli that weigh 10 pounds and the other shows combinations of zucchini and broccoli that cost $17. (Lesson 4-10)

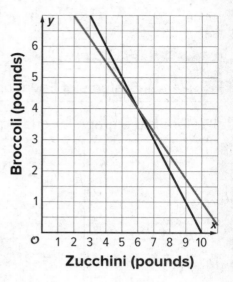

a. Name one combination of veggies that weighs 10 pounds but does not cost $17.

b. Name one combination of veggies that costs $17 but does not weigh 10 pounds.

c. How many pounds each of zucchini and broccoli can Andre get so that he spends all $17 and gets 10 pounds of veggies?

4. The temperature in degrees Fahrenheit, F, is related to the temperature in degrees Celsius, C, by the equation $F = \frac{9}{5}C + 32$. (Lesson 4-9)

a. In the Sahara desert, temperatures often reach 50 degrees Celsius. How many degrees Fahrenheit is this?

b. In parts of Alaska, the temperatures can reach -60 degrees Fahrenheit. How many degrees Celsius is this?

c. There is one temperature where the degrees Fahrenheit and degrees Celsius are the same, so that $C = F$. Use the expression from the equation, where F is expressed in terms of C, to solve for this temperature.

Lesson 4-13

Solving Systems of Equations

NAME _____ DATE _____ PERIOD _____

Learning Goal Let's solve systems of equations.

Warm Up
13.1 True or False: Two Lines

Use the lines to decide whether each statement is true or false. Be prepared to explain your reasoning using the lines.

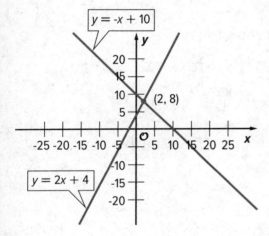

1. A solution to $8 = -x + 10$ is 2.

2. A solution to $2 = 2x + 4$ is 8.

3. A solution to $-x + 10 = 2x + 4$ is 8.

4. A solution to $-x + 10 = 2x + 4$ is 2.

5. There are no values of x and y that make $y = -x + 10$ and $y = 2x + 4$ true at the same time.

Activity

13.2 Matching Graphs to Systems

Here are three **systems of equations** graphed on a coordinate plane.

A
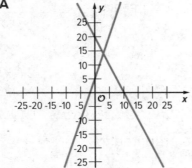

B
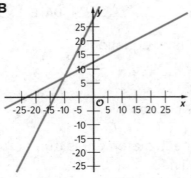

C
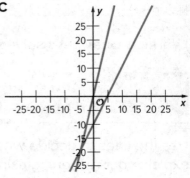

1. Match each figure to one of the systems of equations shown here.

   a. $\begin{cases} y = 3x + 5 \\ y = -2x + 20 \end{cases}$

   b. $\begin{cases} y = 2x - 10 \\ y = 4x - 1 \end{cases}$

   c. $\begin{cases} y = 0.5x + 12 \\ y = 2x + 27 \end{cases}$

2. Find the solution to each system and check that your solution is reasonable based on the graph.

Activity

13.3 Different Types of Systems

Your teacher will give you a page with some systems of equations.

1. Graph each system of equations carefully on the provided coordinate plane.

2. Describe what the graph of a system of equations looks like when it has . . .

   a. 1 solution

   b. 0 solutions

   c. infinitely many solutions

NAME _____ DATE _____ PERIOD _____

The graphs of the equations $Ax + By = 15$ and $Ax - By = 9$ intersect at $(2, 1)$. Find A and B. Show or explain your reasoning.

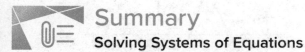

Summary
Solving Systems of Equations

Sometimes it is easier to solve a system of equations without having to graph the equations and look for an intersection point. In general, whenever we are solving a system of equations written as. . .

$$\begin{cases} y = \text{[some stuff]} \\ y = \text{[some other stuff]} \end{cases}$$

we know that we are looking for a pair of values (x, y) that makes both equations true. In particular, we know that the value for y will be the same in both equations. That means that. . .

$$\text{[some stuff]} = \text{[some other stuff]}$$

For example, look at this system of equations:

$$\begin{cases} y = 2x + 6 \\ y = -3x - 4 \end{cases}$$

Since the y value of the solution is the same in both equations, then we know

$$2x + 6 = -3x - 4$$

We can solve this equation for x:

$$
\begin{array}{ll}
2x + 6 = -3x - 4 & \\
5x + 6 = -4 & \text{Add } 3x \text{ to each side.} \\
5x = -10 & \text{Subtract 6 from each side.} \\
x = -2 & \text{Divide each side by 5.}
\end{array}
$$

But this is only half of what we are looking for: we know the value for x, but we need the corresponding value for y. Since both equations have the same y-value, we can use either equation to find the y-value:

$$y = 2(-2) + 6$$

Or

$$y = -3(-2) - 4$$

In both cases, we find that $y = 2$. So the solution to the system is (-2, 2). We can verify this by graphing both equations in the coordinate plane.

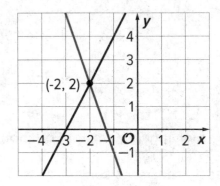

In general, a system of linear equations can have:

- No solutions. In this case, the lines that correspond to each equation never intersect.

- Exactly one solution. The lines that correspond to each equation intersect in exactly one point.

- An infinite number of solutions. The graphs of the two equations are the same line!

NAME _____ DATE _____ PERIOD _____

Practice
Solving Systems of Equations

1. a. Write equations for the lines shown.

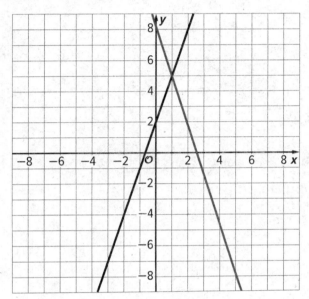

b. Describe how to find the solution to the corresponding system by looking at the graph.

c. Describe how to find the solution to the corresponding system by using the equations.

2. The solution to a system of equations is (5, -19). Choose two equations that might make up the system.

(A.) $y = -3x - 6$

(B.) $y = 2x - 23$

(C.) $y = -7x + 16$

(D.) $y = x - 17$

(E.) $y = -2x - 9$

3. Solve the system of equations: $\begin{cases} y = 4x - 3 \\ y = -2x + 9 \end{cases}$

4. Solve the system of equations: $\begin{cases} y = \frac{5}{4}x - 2 \\ y = \frac{-1}{4}x + 19 \end{cases}$

5. Here is an equation: $\dfrac{15(x - 3)}{5} = 3(2x - 3)$. (Lesson 4-6)

    a. Solve the equation by using the distributive property first.

    b. Solve the equation without using the distributive property.

    c. Check your solution.

Lesson 4-14

Solving More Systems

NAME _____ DATE _____ PERIOD _____

Learning Goal Let's solve systems of equations.

 ## Warm Up
14.1 Algebra Talk: Solving Systems Mentally

Solve these without writing anything down.

1. $\begin{cases} x = 5 \\ y = x - 7 \end{cases}$

2. $\begin{cases} y = 4 \\ y = x + 3 \end{cases}$

3. $\begin{cases} x = 8 \\ y = \text{-}11 \end{cases}$

Here are a lot of systems of equations.

A $\begin{cases} y = 4 \\ x = -5y + 6 \end{cases}$

E $\begin{cases} y = -3x - 5 \\ y = 4x + 30 \end{cases}$

I $\begin{cases} 3x + 4y = 10 \\ x = 2y \end{cases}$

B $\begin{cases} y = 7 \\ x = 3y - 4 \end{cases}$

F $\begin{cases} y = 3x - 2 \\ y = -2x + 8 \end{cases}$

J $\begin{cases} y = 3x + 2 \\ 2x + y = 47 \end{cases}$

C $\begin{cases} y = \frac{3}{2}x + 7 \\ x = -4 \end{cases}$

G $\begin{cases} y = 3x \\ x = -2y + 56 \end{cases}$

K $\begin{cases} y = -2x + 5 \\ 2x + 3y = 31 \end{cases}$

D $\begin{cases} y = -3x + 10 \\ y = -2x + 6 \end{cases}$

H $\begin{cases} x = 2y - 15 \\ y = -2x \end{cases}$

L $\begin{cases} x + y = 10 \\ x = 2y + 1 \end{cases}$

1. Without solving, identify 3 systems that you think would be the least difficult to solve and 3 systems that you think would be the most difficult to solve. Be prepared to explain your reasoning.

2. Choose 4 systems to solve. At least one should be from your "least difficult" list and one should be from your "most difficult" list.

NAME _____ DATE _____ PERIOD _____

Activity

14.3 Five Does Not Equal Seven

Tyler was looking at this system of equations:

$$\begin{cases} x + y = 5 \\ x + y = 7 \end{cases}$$

He said, "Just looking at the system, I can see it has no solution. If you add two numbers, that sum can't be equal to two different numbers."

Do you agree with Tyler?

Are you ready for more?

In rectangle *ABCD*, side *AB* is 8 centimeters and side *BC* is 6 centimeters. *F* is a point on side *BC* and *E* is a point on side *AB*. The area of triangle *DFC* is 20 square centimeters, and the area of triangle *DEF* is 16 square centimeters. What is the area of triangle *AED*?

Summary
Solving More Systems

When we have a system of linear equations where one of the equations is of the form $y =$ [stuff] or $x =$ [stuff], we can solve it algebraically by using a technique called *substitution*. The basic idea is to replace a variable with an expression it is equal to (so the expression is like a substitute for the variable). For example, let's start with the system:

$$\begin{cases} y = 5x \\ 2x - y = 9 \end{cases}$$

Since we know that $y = 5x$, we can substitute $5x$ for y in the equation $2x - y = 9$,

$$2x - (5x) = 9,$$

and then solve the equation for x,

$$x = -3.$$

We can find y using either equation. Using the first one: $y = 5 \cdot -3$. So, $(-3, -15)$ is the solution to this system.

We can verify this by looking at the graphs of the equations in the system:

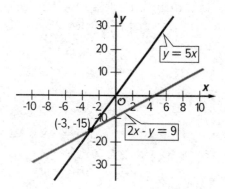

Sure enough! They intersect at $(-3, -15)$.

We didn't know it at the time, but we were actually using substitution in the last lesson as well. In that lesson, we looked at the system

$$\begin{cases} y = 2x + 6 \\ y = -3x - 4 \end{cases}$$

and we substituted $2x + 6$ for y into the second equation to get $2x + 6 = -3x - 4$. Go back and check for yourself!

NAME _____ DATE _____ PERIOD _____

Practice

Solving More Systems

---

1. Solve: $\begin{cases} y = 6x \\ 4x + y = 7 \end{cases}$

2. Solve: $\begin{cases} y = 3x \\ x = -2y + 70 \end{cases}$

3. Which equation, together with $y = -1.5x + 3$, makes a system with one solution?

(A.) $y = -1.5x + 6$

(B.) $y = -1.5x$

(C.) $2y = -3x + 6$

(D.) $2y + 3x = 6$

(E.) $y = -2x + 3$

4. The system $x - 6y = 4$, $3x - 18y = 4$ has no solution.

   a. Change one constant or coefficient to make a new system with one solution.

   b. Change one constant or coefficient to make a new system with an infinite number of solutions.

5. Match each graph to its equation. (Lesson 3-11)

a. $y = 2x + 3$

b. $y = -2x + 3$

c. $y = 2x - 3$

d. $y = -2x - 3$

Graph A

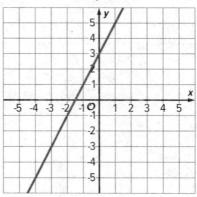

Graph B

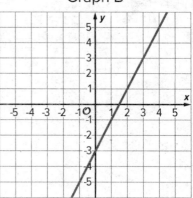

Graph C

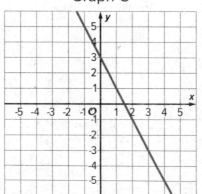

Graph D

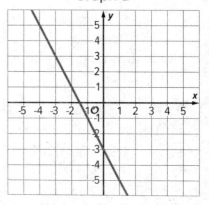

6. Here are two points: (-3, 4), (1, 7). What is the slope of the line between them? (Lesson 3-10)

(A.) $\frac{4}{3}$

(B.) $\frac{3}{4}$

(C.) $\frac{1}{6}$

(D.) $\frac{2}{3}$

Lesson 4-15

Writing Systems of Equations

NAME _____ DATE _____ PERIOD _____

Learning Goal Let's write systems of equations from real-world situations.

 ## Warm Up
15.1 How Many Solutions? Matching

Match each system of equations with the number of solutions the system has.

1. $\begin{cases} y = -\frac{4}{3}x + 4 \\ y = -\frac{4}{3}x - 1 \end{cases}$

2. $\begin{cases} y = 4x - 5 \\ y = -2x + 7 \end{cases}$

3. $\begin{cases} 2x + 3y = 8 \\ 4x + 6y = 17 \end{cases}$

4. $\begin{cases} y = 5x - 15 \\ y = 5(x - 3) \end{cases}$

A. No solutions

B. One solution

C. Infinitely many solutions

 ## Activity
15.2 Situations and Systems

For each situation:

• Create a system of equations.

• Then, without solving, interpret what the solution to the system would tell you about the situation.

1. Lin's family is out for a bike ride when her dad stops to take a picture of the scenery. He tells the rest of the family to keep going and that he'll catch up. Lin's dad spends 5 minutes taking the photo and then rides at 0.24 miles per minute until he meets up with the rest of the family further along the bike path. Lin and the rest were riding at 0.18 miles per minute.

2. Noah is planning a kayaking trip. Kayak Rental A charges a base fee of $15 plus $4.50 per hour. Kayak Rental B charges a base fee of $12.50 plus $5 per hour.

3. Diego is making a large batch of pastries. The recipe calls for 3 strawberries for every apple. Diego used 52 fruits all together.

4. Flour costs $0.80 per pound and sugar costs $0.50 per pound. An order of flour and sugar weighs 15 pounds and costs $9.00.

 Activity

15.3 Info Gap: Racing and Play Tickets

Your teacher will give you either a *problem card* or a *data card*. Do not show or read your card to your partner.

If your teacher gives you the *problem card* . . .	If your teacher gives you the *data card* . . .
1. Silently read your card and think about what information you need to be able to answer the question.	1. Silently read your card.
2. Ask your partner for the specific information that you need.	2. Ask your partner *"What specific information do you need?"* and wait for them to ask for information. If your partner asks for information that is not on the card, do not do the calculations for them. Tell them you don't have that information.
3. Explain how you are using the information to solve the problem. Continue to ask questions until you have enough information to solve the problem.	3. Before sharing the information, ask *"Why do you need that information?"* Listen to your partner's reasoning and ask clarifying questions.
4. Share the *problem card* and solve the problem independently.	4. Read the problem card and solve the problem independently.
5. Read the *data card* and discuss your reasoning.	5. Share the *data card* and discuss your reasoning.

NAME _____ DATE _____ PERIOD _____

Activity
15.4 Solving Systems Practice

Here are a lot of systems of equations.

A $\begin{cases} y = -2x + 6 \\ y = x - 3 \end{cases}$

B $\begin{cases} y = 5x - 4 \\ y = 4x + 12 \end{cases}$

C $\begin{cases} y = \frac{2}{3}x - 4 \\ y = -\frac{4}{3}x + 9 \end{cases}$

D $\begin{cases} 4y + 7x = 6 \\ 4y + 7x = -5 \end{cases}$

E $\begin{cases} y = x - 6 \\ x = 6 + y \end{cases}$

F $\begin{cases} y = 0.24x \\ y = 0.18x + 0.9 \end{cases}$

G $\begin{cases} y = 4.5x + 15 \\ y = 5x + 12.5 \end{cases}$

H $\begin{cases} y = 3x \\ x + y = 52 \end{cases}$

1. Without solving, identify 3 systems that you think would be the least difficult for you to solve and 3 systems you think would be the most difficult. Be prepared to explain your reasoning.

2. Choose 4 systems to solve. At least one should be from your "least difficult" list and one should be from your "most difficult" list.

Summary
Writing Systems of Equations

We have learned how to solve many kinds of systems of equations using algebra that would be difficult to solve by graphing. For example, look at . . .

$$\begin{cases} y = 2x - 3 \\ x + 2y = 7 \end{cases}$$

The first equation says that $y = 2x - 3$, so wherever we see y, we can substitute the expression $2x - 3$ instead. So the second equation becomes $x + 2(2x - 3) = 7$.

We can then solve for x.

$x + 4x - 6 = 7$	Distributive property
$5x - 6 = 7$	Combine like terms.
$5x = 13$	Add 6 to each side.
$x = \dfrac{13}{5}$	Multiply each side by $\dfrac{1}{5}$.

We know that the y-value for the solution is the same for either equation, so we can use either equation to solve for it. Using the first equation, we get . . .

$y = 2\left(\dfrac{13}{5}\right) - 3$	Substitute $x = \dfrac{13}{5}$ into the equation.
$y = \left(\dfrac{26}{5}\right) - 3$	Multiply $2\left(\dfrac{13}{5}\right)$ to make $\left(\dfrac{26}{5}\right)$.
$y = \left(\dfrac{26}{5}\right) - \left(\dfrac{15}{5}\right)$	Rewrite 3 as $\left(\dfrac{15}{5}\right)$.
$y = \dfrac{11}{5}$	

If we substitute $x = \dfrac{13}{5}$ into the other equation, $x + 2y = 7$, we get the same y-value. So the solution to the system is $\left(\dfrac{13}{5}, \dfrac{11}{5}\right)$.

There are many kinds of systems of equations that we will learn how to solve in future grades, like $\begin{cases} 2x + 3y = 6 \\ -x + 2y = 3 \end{cases}$ or even $\begin{cases} y = x^2 + 1 \\ y = 2x + 3 \end{cases}$.

NAME _____ DATE _____ PERIOD _____

Practice
Writing Systems of Equations

1. Kiran and his cousin work during the summer for a landscaping company. Kiran's cousin has been working for the company longer, so his pay is 30% more than Kiran's. Last week his cousin worked 27 hours, and Kiran worked 23 hours. Together, they earned $493.85. What is Kiran's hourly pay? Explain or show your reasoning.

2. Decide which story can be represented by the system of equations $y = x + 6$ and $x + y = 100$. Explain your reasoning.

   a. Diego's teacher writes a test worth 100 points. There are 6 more multiple choice questions than short answer questions.

   b. Lin and her younger cousin measure their heights. They notice that Lin is 6 inches taller, and their heights add up to exactly 100 inches.

3. Clare and Noah play a game in which they earn the same number of points for each goal and lose the same number of points for each penalty. Clare makes 6 goals and 3 penalties, ending the game with 6 points. Noah earns 8 goals and 9 penalties and ends the game with -22 points.

   a. Write a system of equations that describes Clare and Noah's outcomes. Use x to represent the number of points for a goal and y to represent the number of points for a penalty.

   b. Solve the system. What does your solution mean?

4. Solve: $\begin{cases} y = 6x - 8 \\ y = -3x + 10 \end{cases}$ (Lesson 4-14)

5. Respond to each of the following. (Lesson 4-13)

   a. Estimate the coordinates of the point where the two lines meet.

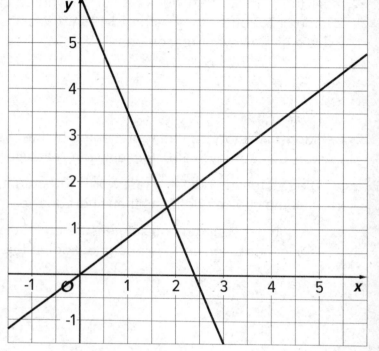

   b. Choose two equations that make up the system represented by the graph.
      - $y = \frac{5}{4}x$
      - $y = 6 - 2.5x$
      - $y = 2.5x + 6$
      - $y = 6 - 3x$
      - $y = 0.8x$

   c. Solve the system of equations and confirm the accuracy of your estimate.

Lesson 4-16

Solving Problems with Systems of Equations

NAME _____ DATE _____ PERIOD _____

Learning Goal Let's solve some gnarly problems.

Warm Up
16.1 Are We There Yet?

A car is driving towards home at 0.5 miles per minute. If the car is 4 miles from home at $t = 0$, which of the following can represent the distance that the car has left to drive?

- $0.5t$

- $4 + 0.5t$

- $4 - 0.5t$

- $4 \cdot (0.5t)$

Activity
16.2 Cycling, Fundraising, Working, and _____?

Solve each problem. Explain or show your reasoning.

1. Two friends live 7 miles apart. One Saturday, the two friends set out on their bikes at 8 am and started riding towards each other. One rides at 0.2 miles per minute, and the other rides at 0.15 miles per minute. At what time will the two friends meet?

2. Students are selling grapefruits and nuts for a fundraiser. The grapefruits cost $1 each and a bag of nuts cost $10 each. They sold 100 items and made $307. How many grapefruits did they sell?

3. Jada earns $7 per hour mowing her neighbors' lawns. Andre gets paid $5 per hour for the first hour of babysitting and $8 per hour for any additional hours he babysits. What is the number of hours they both can work so that they get paid the same amount?

4. Pause here so your teacher can review your work. Then, invent another problem that is like one of these, but with different numbers. Solve your problem.

5. Create a visual display that includes:

   • The new problem you wrote, without the solution.

   • Enough work space for someone to show a solution.

6. Trade your display with another group, and solve each other's new problem. Make sure that you explain your solution carefully. Be prepared to share this solution with the class.

7. When the group that got the problem you invented shares their solution, check that their answer is correct.

Are you ready for more?

On a different Saturday, two friends set out on bikes at 8:00 am and met up at 8:30 am. (The same two friends who live 7 miles apart.) If one was riding at 10 miles per hour, how fast was the other riding?

Learning Targets

Lesson	Learning Target(s)
4-1 Number Puzzles	• I can solve puzzle problems using diagrams, equations, or other representations.
4-2 Keeping the Equation Balanced	• I can add or remove blocks from a hanger and keep the hanger balanced. • I can represent balanced hangers with equations.
4-3 Balanced Moves	• I can add, subtract, multiply, or divide each side of an equation by the same expression to get a new equation with the same solution.
4-4 More Balanced Moves	• I can make sense of multiple ways to solve an equation.

(continued on the next page)

(continued from the previous page)

Lesson	Learning Target(s)
4-5 Solving Any Linear Equation	• I can solve an equation where the variable appears on both sides.
4-6 Strategic Solving	• I can solve linear equations in one variable.
4-7 All, Some, or No Solutions	• I can determine whether an equation has no solutions, one solution, or infinitely many solutions.
4-8 How Many Solutions?	• I can solve equations with different numbers of solutions.

Lesson	Learning Target(s)
4-9 When Are They the Same?	• I can use an expression to find when two things, like height, are the same in a real-world situation.
4-10 On or Off the Line?	• I can identify ordered pairs that are solutions to an equation. • I can interpret ordered pairs that are solutions to an equation.
4-11 On Both of the Lines	• I can use graphs to find an ordered pair that two real-world situations have in common.
4-12 Systems of Equations	• I can explain the solution to a system of equations in a real-world context. • I can explain what a system of equations is. • I can make graphs to find an ordered pair that two real-world situations have in common.

(continued on the next page)

(continued from the previous page)

Lesson	Learning Target(s)
4-13 Solving Systems of Equations	• I can graph a system of equations. • I can solve systems of equations using algebra.
4-14 Solving More Systems	• I can use the structure of equations to help me figure out how many solutions a system of equations has.
4-15 Writing Systems of Equations	• I can write a system of equations from a real-world situation.
4-16 Solving Problems with Systems of Equations	• I can use a system of equations to represent a real-world situation and answer questions about the situation.

Notes:

_____

_____

_____

_____

Unit 5

Functions and Volume

At the end of this unit, you'll apply what you learned about functions and volume to compare the volumes of cylinders, cones, and spheres.

Topics
- Inputs and Outputs
- Representing and Interpreting Functions
- Linear Functions and Rates of Change
- Cylinders and Cones
- Dimensions and Spheres
- Let's Put It to Work

Unit 5
Functions and Volume

Lesson 5-1

Inputs and Outputs

NAME _____ DATE _____ PERIOD _____

Learning Goal Let's make some rules.

Warm Up
1.1 Dividing by 0

Study the statements carefully.

- $12 \div 3 = 4$ because $12 = 4 \cdot 3$
- $6 \div 0 = x$ because $6 = x \cdot 0$

What value can be used in place of x to create true statements?
Explain your reasoning.

Activity
1.2 Guess My Rule

Keep the rule cards face down. Decide who will go first.

1. Player 1 picks up a card and silently reads the rule without showing it to Player 2.

2. Player 2 chooses an integer and asks Player 1 for the result of applying the rule to that number.

3. Player 1 gives the result, without saying how they got it.

4. Keep going until Player 2 correctly guesses the rule.

After each round, the players switch roles.

If you have a rule, you can apply it several times in a row and look for patterns. For example, if your rule was "add 1" and you started with the number 5, then by applying that rule over and over again you would get 6, then 7, then 8, etc., forming an obvious pattern.

Try this for the rules in this activity. That is, start with the number 5 and apply each of the rules a few times. Do you notice any patterns? What if you start with a different starting number?

Activity

1.3 Making Tables

For each input-output rule, fill in the table with the outputs that go with a given input. Add two more input-output pairs to the table.

1.

$\frac{3}{4}$ → add 1 then multiply by 4 → 7

Input	Output
$\frac{3}{4}$	7
2.35	
42	

2.

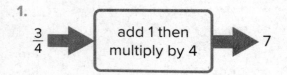

$\frac{3}{4}$ → name the digit in the tenths place → 7

Input	Output
$\frac{3}{4}$	7
2.35	
42	

NAME _____ DATE _____ PERIOD _____

3.

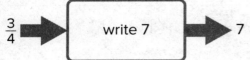

$\frac{3}{4}$ ➡ write 7 ➡ 7

Input	Output
$\frac{3}{4}$	7
2.35	
42	

Pause here until your teacher directs you to the last rule.

4.

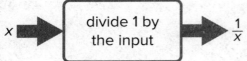

x ➡ divide 1 by the input ➡ $\frac{1}{x}$

Input	Output
$\frac{3}{7}$	$\frac{7}{3}$
1	
0	

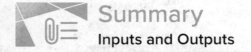

Summary
Inputs and Outputs

An *input-output rule* is a rule that takes an allowable input and uses it to determine an output.

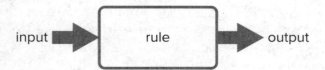

input ➡️ | rule | ➡️ output

For example, the following diagram represents the rule that takes any number as an input, then adds 1, multiplies by 4, and gives the resulting number as an output.

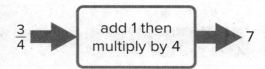

$\frac{3}{4}$ ➡️ | add 1 then multiply by 4 | ➡️ 7

In some cases, not all inputs are allowable, and the rule must specify which inputs will work. For example, this rule is fine when the input is 2:

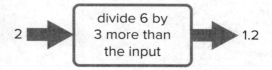

2 ➡️ | divide 6 by 3 more than the input | ➡️ 1.2

But if the input is -3, we would need to evaluate 6 ÷ 0 to get the output.

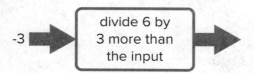

-3 ➡️ | divide 6 by 3 more than the input | ➡️

So, when we say that the rule is "divide 6 by 3 more than the input," we also have to say that -3 is not allowed as an input.

NAME _____ DATE _____ PERIOD _____

Practice
Inputs and Outputs

1. Given the rule:

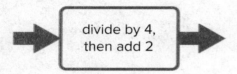

divide by 4, then add 2

Complete the table for the function rule for the following input values:

Input	0	2	4	6	8	10
Output						

2. Here is an input-output rule:

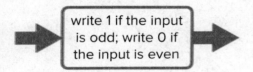

write 1 if the input is odd; write 0 if the input is even

Complete the table for the input-output rule:

Input	-3	-2	-1	0	1	2	3
Output							

3. Andre's school orders some new supplies for the chemistry lab. The online store shows a pack of 10 test tubes costs $4 less than a set of nested beakers. In order to fully equip the lab, the school orders 12 sets of beakers and 8 packs of test tubes. **(Lesson 4-15)**

a. Write an equation that shows the cost of a pack of test tubes, t, in terms of the cost of a set of beakers, b.

b. The school office receives a bill for the supplies in the amount of $348. Write an equation with t and b that described this situation.

c. Since t is in terms of b from the first equation, this expression can be substituted into the second equation where t appears. Write an equation that shows this substitution.

d. Solve the equation for b.

e. How much did the school pay for a set of beakers? For a pack of test tubes?

4. Solve: $\begin{cases} y = x - 4 \\ y = 6x - 10 \end{cases}$ (Lesson 4-14)

5. For what value of x do the expressions $2x + 3$ and $3x - 6$ have the same value? (Lesson 4-9)

Lesson 5-2

Introduction to Functions

NAME _____ DATE _____ PERIOD _____

Learning Goal Let's learn what a function is.

Warm Up
2.1 Square Me

Here are some numbers in a list.

$$1, \ -3, \ -\frac{1}{2}, \ 3, \ 2, \ \frac{1}{4}, \ 0.5$$

1. How many different numbers are in the list?

2. Make a new list containing the squares of all these numbers.

3. How many different numbers are in the new list?

4. Explain why the two lists do not have the same number of different numbers.

Activity

2.2 You Know This, Do You Know That?

Say yes or no for each question. If yes, draw an input-output diagram. If no, give examples of two different outputs that are possible for the same input.

1. A person is 5.5 feet tall. Do you know their height in inches?

2. A number is 5. Do you know its square?

3. The square of a number is 16. Do you know the number?

4. A square has a perimeter of 12 cm. Do you know its area?

5. A rectangle has an area of 16 cm^2. Do you know its length?

6. You are given a number. Do you know the number that is $\frac{1}{5}$ as big?

7. You are given a number. Do you know its reciprocal?

NAME _____   DATE _____   PERIOD _____

Activity
2.3 Using Function Language

Here are the questions from the previous activity. For the ones you said yes to, write a statement like, "The height a rubber ball bounces to depends on the height it was dropped from" or "Bounce height is a **function** of drop height." For all of the ones you said no to, write a statement like, "The day of the week does not determine the temperature that day" or "The temperature that day is not a function of the day of the week."

1. A person is 5.5 feet tall. Do you know their height in inches?

2. A number is 5. Do you know its square?

3. The square of a number is 16. Do you know the number?

4. A square has a perimeter of 12 cm. Do you know its area?

5. A rectangle has an area of 16 cm². Do you know its length?

6. You are given a number. Do you know the number that is $\frac{1}{5}$ as big?

7. You are given a number. Do you know its reciprocal?

Activity

2.4 Same Function, Different Rule?

Which input-output rules could describe the same function (if any)? Be prepared to explain your reasoning.

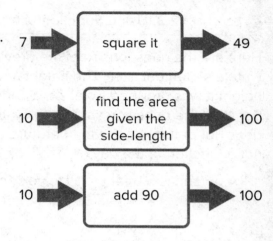

Are you ready for more?

The phrase "is a function of" gets used in non-mathematical speech as well as mathematical speech in sentences like, "The range of foods you like is a function of your upbringing." What is that sentence trying to convey? Is it the same use of the word "function" as the mathematical one?

Summary

Introduction to Functions

Let's say we have an input-output rule that for each allowable input gives exactly one output. Then we say the output *depends* on the input, or the output is a **function** of the input.

For example, the area of a square is a function of the side length, because you can find the area from the side length by squaring it. So when the input is 10 cm, the output is 100 cm².

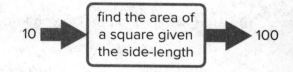

Sometimes we might have two different rules that describe the same function. As long as we always get the same, single output from the same input, the rules describe the same function.

Glossary

function

NAME _____ DATE _____ PERIOD _____

Practice
Introduction to Functions

1. Here are several function rules. Calculate the output for each rule when you use -6 as the input.

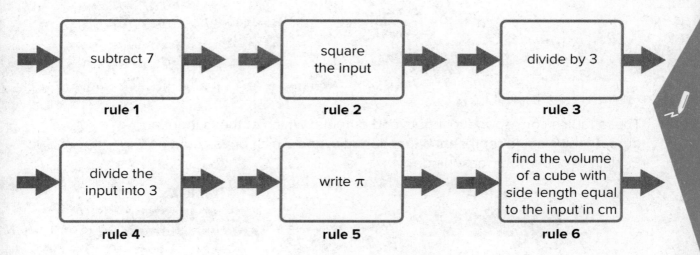

2. A group of students is timed while sprinting 100 meters. Each student's speed can be found by dividing 100 m by their time. Is each statement true or false? Explain your reasoning.

a. Speed is a function of time.

b. Time is a function of distance.

c. Speed is a function of number of students racing.

d. Time is a function of speed.

3. Diego's history teacher writes a test for the class with 26 questions. The test is worth 123 points and has two types of questions: multiple choice worth 3 points each, and essays worth 8 points each. How many essay questions are on the test? Explain or show your reasoning. (Lesson 4-15)

4. These tables correspond to inputs and outputs. Which of these input and output tables could represent a function rule, and which ones could not? Explain or show your reasoning.

Table A:

Input	Output
-2	4
-1	1
0	0
1	1
2	4

Table B:

Input	Output
4	-2
1	-1
0	0
1	1
4	2

Table C:

Input	Output
1	0
2	0
3	0

Table D:

Input	Output
0	1
0	2
0	3

Lesson 5-3

Equations for Functions

NAME _____ DATE _____ PERIOD _____

Learning Goal Let's find outputs from equations.

Warm Up
3.1 A Square's Area

Fill in the table of input-output pairs for the given rule. Write an algebraic expression for the rule in the box in the diagram.

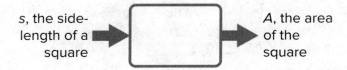

s, the side-length of a square → [ ] → A, the area of the square

Input	Output
8	
2.2	
$12\frac{1}{4}$	
s	

Activity

3.2 Diagrams, Equations, and Descriptions

Record your answers to these questions in the table provided.

1. Match each of these descriptions with a diagram.

   a. the circumference, C, of a circle with **radius**, r

   b. the distance in miles, d, that you would travel in t hours if you drive at 60 miles per hour

   c. the output when you triple the input and subtract 4

   d. the volume of a cube, v given its edge length, s

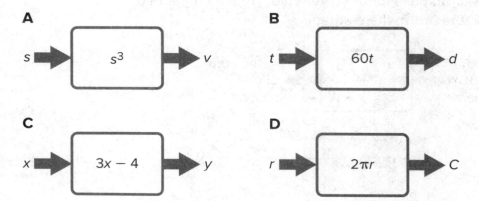

2. Write an equation for each description that expresses the output as a function of the input.

3. Find the output when the input is 5 for each equation.

4. Name the **independent** and **dependent variables** of each equation.

Description	a	b	c	d
diagram				
equation				
input = 5 output = ?				
independent variable				
dependent variable				

NAME _____ DATE _____ PERIOD _____

Are you ready for more?

Choose a 3-digit number as an input and apply the following rule to it, one step at a time:

- Multiply your number by 7.

- Add one to the result.

- Multiply the result by 11.

- Subtract 5 from the result.

- Multiply the result by 13.

- Subtract 78 from the result to get the output.

- Can you describe a simpler way to describe this rule?
  Why does this work?

Activity

3.3 Dimes and Quarters

Jada had some dimes and quarters that had a total value of $12.50. The relationship between the number of dimes, d, and the number of quarters, q, can be expressed by the equation $0.1d + 0.25q = 12.5$.

1. If Jada has 4 quarters, how many dimes does she have?

2. If Jada has 10 quarters, how many dimes does she have?

3. Is the number of dimes a function of the number of quarters? If yes, write a rule (that starts with $d = ...$) that you can use to determine the output, d, from a given input, q. If no, explain why not.

4. If Jada has 25 dimes, how many quarters does she have?

NAME _____ DATE _____ PERIOD _____

5. If Jada has 30 dimes, how many quarters does she have?

6. Is the number of quarters a function of the number of dimes? If yes, write a rule (that starts with $q = ...$) that you can use to determine the output, q, from a given input, d. If no, explain why not.

Summary
Equations for Functions

We can sometimes represent functions with equations.

For example, the area, A, of a circle is a function of the radius, r, and we can express this with an equation:

$$A = \pi r^2$$

We can also draw a diagram to represent this function:

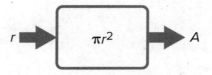

In this case, we think of the radius, r, as the input, and the area of the circle, A, as the output. For example, if the input is a radius of 10 cm, then the output is an area of 100π cm^2, or about 314 square cm. Because this is a function, we can find the area, A, for any given radius, r.

Since it is the input, we say that r is the **independent variable** and, as the output, A is the **dependent variable**.

Sometimes when we have an equation we get to choose which variable is the independent variable.

For example, if we know that ...

$$10A - 4B = 120$$

then we can think of A as a function of B and write ...

$$A = 0.4B + 12$$

or we can think of B as a function of A and write ...

$$B = 2.5A - 30.$$

Glossary

dependent variable
independent variable
radius

NAME _____ DATE _____ PERIOD _____

Practice
Equations for Functions

1. Here is an equation that represents a function: $72x + 12y = 60$.

   Select **all** the different equations that describe the same function:

   (A.) $120y + 720x = 600$ (E.) $x = \frac{5}{6} - \frac{y}{6}$

   (B.) $y = 5 - 6x$ (F.) $7x + 2y = 6$

   (C.) $2y + 12x = 10$ (G.) $x = \frac{5}{6} + \frac{y}{6}$

   (D.) $y = 5 + 6x$

2. Respond to each of the following. **(Lesson 4-13)**

   a. Graph a system of linear equations with no solution.

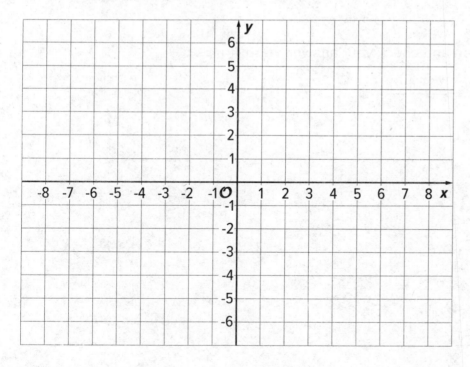

   b. Write an equation for each line you graph.

3. Brown rice costs $2 per pound, and beans cost $1.60 per pound. Lin has $10 to spend on these items to make a large meal of beans and rice for a potluck dinner. Let b be the number of pounds of beans Lin buys and r be the number of pounds of rice she buys when she spends all her money on this meal.

   a. Write an equation relating the two variables.

   b. Rearrange the equation so b is the independent variable.

   c. Rearrange the equation so r is the independent variable.

4. Solve each equation and check your answer. (Lesson 4-6)

   a. $2x + 4(3 - 2x) = \dfrac{3(2x + 2)}{6} + 4$

   b. $4z + 5 = \text{-}3z - 8$

   c. $\dfrac{1}{2} - \dfrac{1}{8}q = \dfrac{q - 1}{4}$

Lesson 5-4

Tables, Equations, and Graphs of Functions

NAME _____ DATE _____ PERIOD _____

Learning Goal Let's connect equations and graphs of functions.

Warm Up
4.1 Notice and Wonder: Doubling Back

What do you notice? What do you wonder?

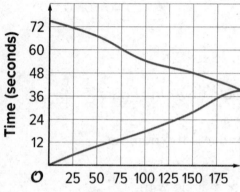

Activity

4.2 Equations and Graphs of Functions

The graphs of three functions are shown.

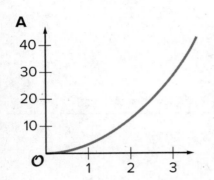

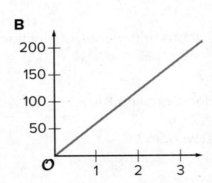

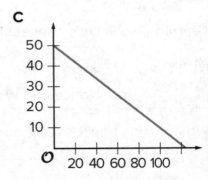

1. Match one of these equations to each of the graphs.

   a. $d = 60t$, where d is the distance in miles that you would travel in t hours if you drove at 60 miles per hour.

   b. $q = 50 - 0.4d$, where q is the number of quarters, and d is the number of dimes, in a pile of coins worth $12.50.

   c. $A = \pi r^2$, where A is the area in square centimeters of a circle with radius r centimeters.

2. Label each of the axes with the independent and dependent variables and the quantities they represent.

3. For each function: What is the output when the input is 1? What does this tell you about the situation? Label the corresponding point on the graph.

4. Find two more input-output pairs. What do they tell you about the situation? Label the corresponding points on the graph.

NAME _____ DATE _____ PERIOD _____

Are you ready for more?

A function inputs fractions $\frac{a}{b}$ between 0 and 1 where a and b have no common factors, and outputs the fraction $\frac{1}{b}$.

For example, given the input $\frac{3}{4}$ the function outputs $\frac{1}{4}$, and to the input $\frac{1}{2}$ the function outputs $\frac{1}{2}$. These two input-output pairs are shown on the graph.

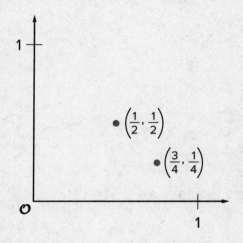

Plot at least 10 more points on the graph of this function.

Are most points on the graph above or below a height of 0.3?
Of height 0.01?

Activity

4.3 Running around a Track

1. Kiran was running around the track. The graph shows the time, t, he took to run various distances, d. The table shows his time in seconds after every three meters.

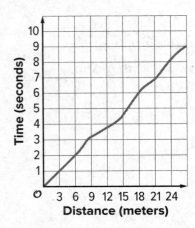

d	0	3	6	9	12	15	18	21	24	27
t	0	1.0	2.0	3.2	3.8	4.6	6.0	6.9	8.09	9.0

a. How long did it take Kiran to run 6 meters?

b. How far had he gone after 6 seconds?

c. Estimate when he had run 19.5 meters.

d. Estimate how far he ran in 4 seconds.

e. Is Kiran's time a function of the distance he has run? Explain how you know.

NAME _____ DATE _____ PERIOD _____

2. Priya is running once around the track. The graph shows her time given how far she is from her starting point.

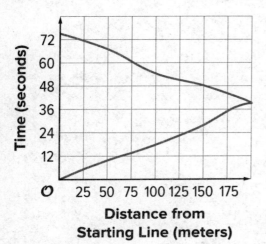

a. What was her farthest distance from her starting point?

b. Estimate how long it took her to run around the track.

c. Estimate when she was 100 meters from her starting point.

d. Estimate how far she was from the starting line after 60 seconds.

e. Is Priya's time a function of her distance from her starting point? Explain how you know.

Here is the graph showing Noah's run.

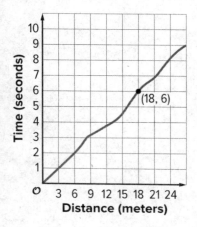

The time in seconds since he started running is a function of the distance he has run. The point (18, 6) on the graph tells you that the time it takes him to run 18 meters is 6 seconds. The input is 18 and the output is 6.

The graph of a function is all the coordinate pairs, (input, output), plotted in the coordinate plane. By convention, we always put the input first, which means that the inputs are represented on the horizontal axis and the outputs, on the vertical axis.

NAME _____ DATE _____ PERIOD _____

Practice
Tables, Equations, and Graphs of Functions

1. The graph and the table show the high temperatures in a city over a 10-day period.

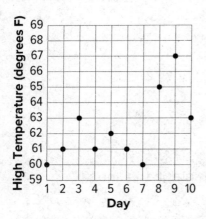

Day	1	2	3	4	5	6	7	8	9	10
Temperature (degrees F)	60	61	63	61	62	61	60	65	67	63

a. What was the high temperature on Day 7?

b. On which days was the high temperature 61 degrees?

c. Is the high temperature a function of the day? Explain how you know.

d. Is the day a function of the high temperature? Explain how you know.

2. The amount Lin's sister earns at her part-time job is proportional to the number of hours she works. She earns $9.60 per hour.

    a. Write an equation in the form $y = kx$ to describe this situation, where x represents the hours she works and y represents the dollars she earns.

    b. Is y a function of x? Explain how you know.

    c. Write an equation describing x as a function of y.

3. Use the equation $2m + 4s = 16$ to complete the table, then graph the line using s as the dependent variable.

m	0		-2	
s		3		0

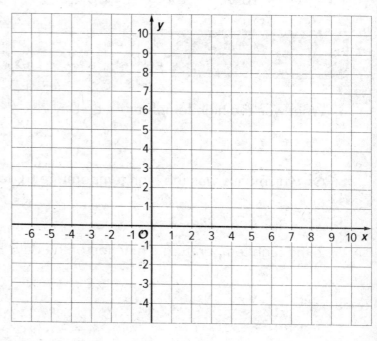

4. Solve the system of equations: $\begin{cases} y = 7x + 10 \\ y = -4x - 23 \end{cases}$ **(Lesson 4-13)**

Lesson 5-5

More Graphs of Functions

NAME _____ DATE _____ PERIOD _____

Learning Goal Let's interpret graphs of functions.

Warm Up
5.1 Which One Doesn't Belong: Graphs

Which graph doesn't belong?

Graph A

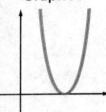

Graph B

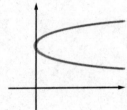

Graph C

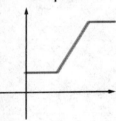

Graph D

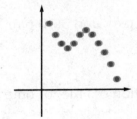

Activity

5.2 Time and Temperature

The graph shows the temperature between noon and midnight in one day in a certain city.

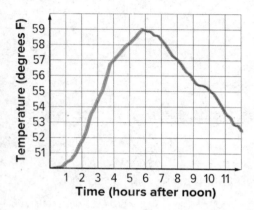

1. Was it warmer at 3:00 p.m. or 9:00 p.m.?

2. Approximately when was the temperature highest?

3. Find another time that the temperature was the same as it was at 4:00 p.m.

4. Did the temperature change more between 1:00 p.m. and 3:00 p.m. or between 3:00 p.m. and 5:00 p.m.?

5. Does this graph show that temperature is a function of time, or time is a function of temperature?

6. When the input for the function is 8, what is the output? What does that tell you about the time and temperature?

NAME _____ DATE _____ PERIOD _____

Activity

5.3 Garbage

1. The graph shows the amount of garbage produced in the US each year between 1991 and 2013.

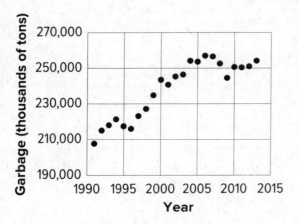

a. Did the amount of garbage increase or decrease between 1999 and 2000?

b. Did the amount of garbage increase or decrease between 2005 and 2009?

c. Between 1991 and 1995, the garbage increased for three years, and then it decreased in the fourth year. Describe how the amount of garbage changed in the years between 1995 and 2000.

2. The graph shows the percentage of garbage that was recycled between 1991 and 2013.

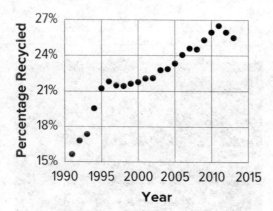

a. When was it increasing?

b. When was it decreasing?

c. Tell the story of the change in the percentage of garbage recycled in the US over this time period.

NAME _____ DATE _____ PERIOD _____

Are you ready for more?

Refer to the graph in the first part of the activity.

1. Find a year where the amount of garbage produced increased from the previous year, but not by as much it increased the following year.

2. Find a year where the amount of garbage produced increased from the previous year, and then increased by a smaller amount the following year.

3. Find a year where the amount of garbage produced decreased from the previous year, but not by as much it decreased the following year.

4. Find a year where the amount of garbage produced decreased from the previous year, and then decreased by a smaller amount the following year.

Here is a graph showing the temperature in a town as a function of time after 8:00 p.m.

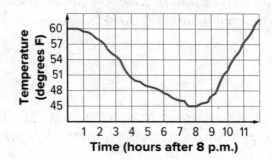

The graph of a function tells us what is happening in the context the function represents.

In this example, the temperature starts out at 60°F at 8:00 p.m. It decreases during the night, reaching its lowest point at 8 hours after 8:00 p.m., or 4:00 a.m. Then it starts to increase again.

NAME _____ DATE _____ PERIOD _____

Practice
More Graphs of Functions

1. The solution to a system of equations is (6, -3). Choose two equations that might make up the system. **(Lesson 4-13)**

   (A.) $y = -3x + 6$

   (B.) $y = 2x - 9$

   (C.) $y = -5x + 27$

   (D.) $y = 2x - 15$

   (E.) $y = -4x + 27$

2. A car is traveling on a small highway and is either going 55 miles per hour or 35 miles per hour, depending on the speed limits, until it reaches its destination 200 miles away. Letting x represent the amount of time in hours that the car is going 55 miles per hour, and y being the time in hours that the car is going 35 miles per hour, an equation describing the relationship is $55x + 35y = 200$. **(Lesson 5-3)**

   a. If the car spends 2.5 hours going 35 miles per hour on the trip, how long does it spend going 55 miles per hour?

   b. If the car spends 3 hours going 55 miles per hour on the trip, how long does it spend going 35 miles per hour?

   c. If the car spends no time going 35 miles per hour, how long would the trip take? Explain your reasoning.

3. The graph represents an object that is shot upwards from a tower and then falls to the ground. The independent variable is time in seconds and the dependent variable is the object's height above the ground in meters.

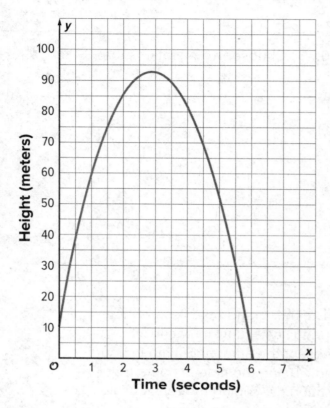

a. How tall is the tower from which the object was shot?

b. When did the object hit the ground?

c. Estimate the greatest height the object reached and the time it took to reach that height. Indicate this situation on the graph.

Lesson 5-6

Even More Graphs of Functions

NAME _____ DATE _____ PERIOD _____

Learning Goal Let's draw a graph from a story.

 **Warm Up**
6.1 Dog Run

Here are five pictures of a dog taken at equal intervals of time.

Diego and Lin drew different graphs to represent this situation:

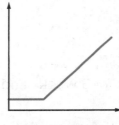 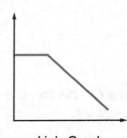

Diego's graph   **Lin's Graph**

They both used time as the independent variable. What do you think each one used for the dependent variable? Explain your reasoning.

For each situation,

- name the independent and dependent variables

- pick the graph that best fits the situation, or sketch the graph if one isn't provided

- label the axes

- answer the question: which quantity is a function of which? Be prepared to explain your reasoning.

1. Jada is training for a swimming race. The more she practices, the less time it takes for her to swim one lap.

2. Andre adds some money to a jar in his room each week for 3 weeks and then takes some out in week 4.

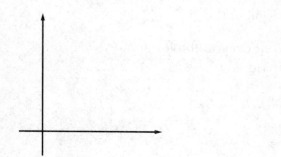

NAME _____ DATE _____ PERIOD _____

 Activity

6.3 Sketching a Story about a Boy and a Bike

Your teacher will give you tools for creating a visual display. With your group, create a display that shows your response to each question.

Here is a story: "Noah was at home. He got on his bike and rode to his friend's house and stayed there for awhile. Then he rode home again. Then he rode to the park. Then he rode home again."

1. Create a set of axes and sketch a graph of this story.

2. What are the two quantities? Label the axes with their names and units of measure. (For example, if this were a story about pouring water into a pitcher, one of your labels might say "volume (liters).")

3. Which quantity is a function of which? Explain your reasoning.

4. Based on your graph, is his friend's house or the park closer to Noah's home? Explain how you know.

5. Read the story and all your responses again. Does everything make sense? If not, make changes to your work.

It is the year 3000. Noah's descendants are still racing around the park, but thanks to incredible technological advances, now with much more powerful gadgets at their disposal. How might their newfound access to teleportation and time-travel devices alter the graph of stories of their daily adventures? Could they affect whether or not the distance from home is a function of the time elapsed?

Summary
Even More Graphs of Functions

Here is a graph showing Andre's distance as a function of time.

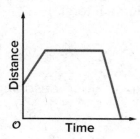

For a graph representing a context, it is important to specify the quantities represented on each axis.

For example, if this is showing distance from home, then Andre starts at some distance from home (maybe at his friend's house), moves further away (maybe to a park), then returns home.

If instead the graph is showing distance from school, the story may be Andre starts out at home, moves further away (maybe to a friend's house), then goes to school.

What could the story be if the graph is showing distance from a park?

NAME _____ DATE _____ PERIOD _____

Practice
Even More Graphs of Functions

1. Match the graph to the following situations (you can use a graph multiple times). For each match, name possible independent and dependent variables and how you would label the axes.

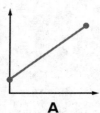
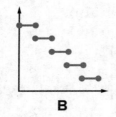
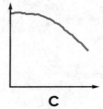

A **B** **C**

a. Tyler pours the same amount of milk from a bottle every morning.

b. A plant grows the same amount every week.

c. The day started very warm but then it got colder.

d. A carnival has an entry fee of $5 and tickets for rides cost $1 each.

2. Jada fills her aquarium with water. The graph shows the height of the water, in cm, in the aquarium as a function of time in minutes. Invent a story of how Jada fills the aquarium that fits the graph.

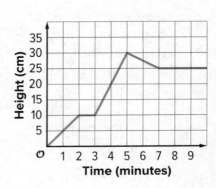

3. Recall the formula for area of a circle. (Lesson 5-4)

   a. Write an equation relating a circle's radius, r, and area, A.

   b. Is area a function of the radius? Is radius a function of the area?

   c. Fill in the missing parts of the table.

r	3		$\frac{1}{2}$	
A		16π		100π

4. The points with coordinates (4, 8), (2, 10), and (5, 7) all lie on the line $2x + 2y = 24$. (Lesson 3-11)

   a. Create a graph, plot the points, and sketch the line.

   b. What is the slope of the line you graphed?

   c. What does this slope tell you about the relationship between lengths and widths of rectangles with perimeter 24?

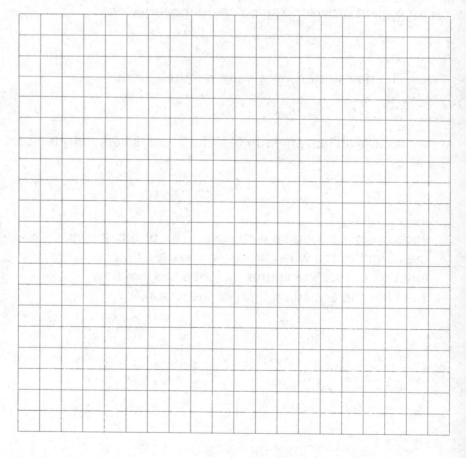

Lesson 5-7

Connecting Representations of Functions

NAME _____ DATE _____ PERIOD _____

Learning Goal Let's connect tables, equations, graphs, and stories of functions.

Warm Up
7.1 Which Are the Same? Which Are Different?

Here are three different ways of representing functions. How are they alike? How are they different?

1. $y = 2x$

2.

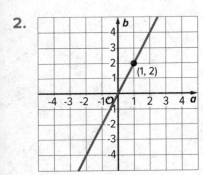

3.

p	-2	-1	0	1	2	3
q	4	2	0	-2	-4	-6

The graph shows the temperature between noon and midnight in City A on a certain day.

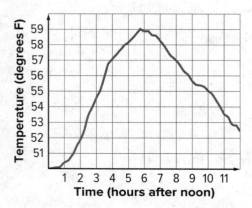

The table shows the temperature, T, in degrees Fahrenheit, for h hours after noon, in City B.

h	1	2	3	4	5	6
T	82	78	75	62	58	59

1. Which city was warmer at 4:00 p.m.?

2. Which city had a bigger change in temperature between 1:00 p.m. and 5:00 p.m.?

3. How much greater was the highest recorded temperature in City B than the highest recorded temperature in City A during this time?

4. Compare the outputs of the functions when the input is 3.

NAME _____ DATE _____ PERIOD _____

Activity

7.3 Comparing Volumes

The **volume**, V, of a cube with edge length s cm is given by the equation $V = s^3$.

The volume of a sphere is a function of its radius (in centimeters), and the graph of this relationship is shown here.

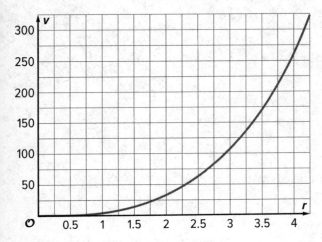

1. Is the volume of a cube with edge length $s = 3$ greater or less than the volume of a sphere with radius 3?

2. If a sphere has the same volume as a cube with edge length 5, estimate the radius of the sphere.

3. Compare the outputs of the two volume functions when the inputs are 2.

Are you ready for more?

Estimate the edge length of a cube that has the same volume as a sphere with radius 2.5.

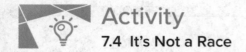

Activity

7.4 It's Not a Race

Elena's family is driving on the freeway at 55 miles per hour.

Andre's family is driving on the same freeway, but not at a constant speed. The table shows how far Andre's family has traveled, d, in miles, every minute for 10 minutes.

t	1	2	3	4	5	6	7	8	9	10
d	0.9	1.9	3.0	4.1	5.1	6.2	6.8	7.4	8	9.1

1. How many miles per minute is 55 miles per hour?

2. Who had traveled farther after 5 minutes? After 10 minutes?

3. How long did it take Elena's family to travel as far as Andre's family had traveled after 8 minutes?

4. For both families, the distance in miles is a function of time in minutes. Compare the outputs of these functions when the input is 3.

NAME _____ DATE _____ PERIOD _____

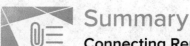

Summary

Connecting Representations of Functions

Functions are all about getting outputs from inputs. For each way of representing a function—equation, graph, table, or verbal description—we can determine the output for a given input.

Let's say we have a function represented by the equation $y = 3x + 2$ where y is the dependent variable and x is the independent variable. If we wanted to find the output that goes with 2, we can input 2 into the equation for x and finding the corresponding value of y. In this case, when x is 2, y is 8 since $3 \cdot 2 + 2 = 8$.

If we had a graph of this function instead, then the coordinates of points on the graph are the input-output pairs. So we would read the y-coordinate of the point on the graph that corresponds to a value of 2 for x. Looking at the graph of this function here, we can see the point (2, 8) on it, so the output is 8 when the input is 2.

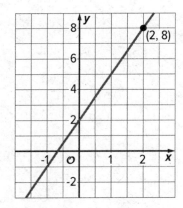

A table representing this function shows the input-output pairs directly (although only for select inputs).

x	-1	0	1	2	3
y	-1	2	5	8	11

Again, the table shows that if the input is 2, the output is 8.

Glossary

volume

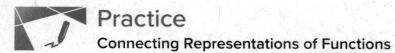
1. The equation and the tables represent two different functions. Use the equation $b = 4a - 5$ and the table to answer the questions. This table represents c as a function of a.

a	-3	0	2	5	10	12
c	-20	7	3	21	19	45

a. When a is -3, is b or c greater?

b. When c is 21, what is the value of a? What is the value of b that goes with this value of a?

c. When a is 6, is b or c greater?

d. For what values of a do we know that c is greater than b?

NAME _____ DATE _____ PERIOD _____

2. Elena and Lin are training for a race. Elena runs her mile at a constant speed of 7.5 miles per hour.

Lin's times are recorded every minute:

Time (minutes)	1	2	3	4	5	6	7	8	9
Distance (miles)	0.11	0.21	0.32	0.41	0.53	0.62	0.73	0.85	1

a. Who finished their mile first?

b. This is a graph of Lin's progress. Draw a graph to represent Elena's mile on the same axes.

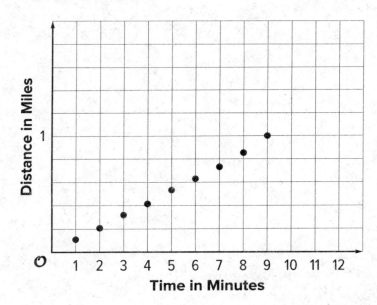

c. For these models, is distance a function of time? Is time a function of distance? Explain how you know.

3. Match each function rule with the value that could not be a possible input for that function. (Lesson 5-2)

    a. 3 divided by the input

                                                      i. 3

    b. Add 4 to the input, then divide this value into 3

                                                   ii. 4

                                                   iii. -4

    c. Subtract 3 from the input, then divide this value into 1

                                                   iv. 0

                                                   v. 1

4. Find a value of x that makes the equation true:

$-(-2x + 1) = 9 - 14x$

Explain your reasoning, and check that your answer is correct. (Lesson 4-4)

Lesson 5-8

Linear Functions

NAME _____ DATE _____ PERIOD _____

Learning Goal Let's investigate linear functions.

Warm Up
8.1 Bigger and Smaller

Diego said that these graphs are ordered from smallest to largest. Mai said they are ordered from largest to smallest. But these are graphs, not numbers! What do you think Diego and Mai are thinking?

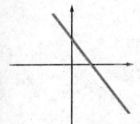 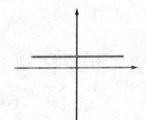 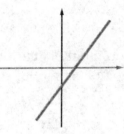

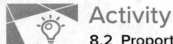

8.2 Proportional Relationships Define Linear Functions

1. Jada earns $7 per hour mowing her neighbors' lawns.

    a. Name the two quantities in this situation that are in a functional relationship. Which did you choose to be the independent variable? What is the variable that depends on it?

    b. Write an equation that represents the function.

    c. Here is a graph of the function. Label the axes. Label at least two points with input-output pairs.

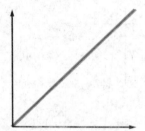

2. To convert feet to yards, you multiply the number of feet by $\frac{1}{3}$.

    a. Name the two quantities in this situation that are in a functional relationship. Which did you choose to be the independent variable? What is the variable that depends on it?

    b. Write an equation that represents the function.

NAME _____ DATE _____ PERIOD _____

c. Draw the graph of the function. Label at least two points with input-output pairs.

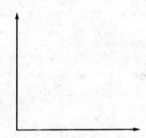
Activity
8.3 Is It Filling Up or Draining Out?

There are four tanks of water. The amount of water in gallons, A, in Tank A is given by the function $A = 200 + 8t$, where t is in minutes. The amount of water in gallons, B, in Tank B starts at 400 gallons and is decreasing at 5 gallons per minute. These functions work when $t \geq 0$ and $t \leq 80$.

1. Which tank started out with more water?

2. Write an equation representing the relationship between B and t.

3. One tank is filling up. The other is draining out. Which is which? How can you tell?

4. The amount of water in gallons, C, in Tank C is given by the function $C = 800 - 7t$. Is it filling up or draining out? Can you tell just by looking at the equation?

5. The graph of the function for the amount of water in gallons, D, in Tank D at time t is shown. Is it filling up or draining out? How do you know?

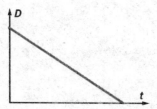

Are you ready for more?

- Pick a tank that was draining out. How long did it take for that tank to drain? What percent full was the tank when 30% of that time had elapsed? When 70% of the time had elapsed?

- What point in the plane is 30% of the way from $(0, 15)$ to $(5, 0)$? 70% of the way?

- What point in the plane is 30% of the way from $(3, 5)$ to $(8, 6)$? 70% of the way?

NAME _____ DATE _____ PERIOD _____

Activity
8.4 Which Is Growing Faster?

Noah is depositing money in his account every week to save money. The graph shows the amount he has saved as a function of time since he opened his account.

Elena opened an account the same day as Noah. The amount of money E in her account is given by the function $E = 8w + 60$, where w is the number of weeks since the account was opened.

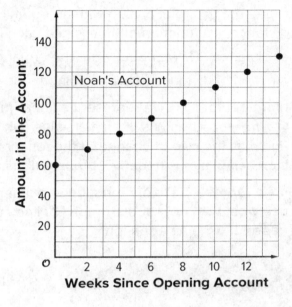

1. Who started out with more money in their account? Explain how you know.

2. Who is saving money at a faster rate? Explain how you know.

3. How much will Noah save over the course of a year if he does not make any withdrawals? How long will it take Elena to save that much?

Summary
Linear Functions

Suppose a car is traveling at 30 miles per hour. The relationship between the time in hours and the distance in miles is a proportional relationship. We can represent this relationship with an equation of the form $d = 30t$, where distance is a function of time (since each input of time has exactly one output of distance). Or we could write the equation $t = \frac{1}{30}d$ instead, where time is a function of distance (since each input of distance has exactly one output of time).

More generally, if we represent a linear function with an equation like $y = mx + b$, then b is the initial value (which is 0 for proportional relationships), and m is the rate of change of the function. If m is positive, the function is increasing. If m is negative, the function is decreasing.

If we represent a linear function in a different way, say with a graph, we can use what we know about graphs of lines to find the m and b values and, if needed, write an equation.

NAME _____ DATE _____ PERIOD _____

Practice
Linear Functions

1. Two cars drive on the same highway in the same direction. The graphs show the distance, *d*, of each one as a function of time, *t*. Which car drives faster? Explain how you know.

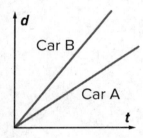

2. Two car services offer to pick you up and take you to your destination. Service A charges 40 cents to pick you up and 30 cents for each mile of your trip. Service B charges $1.10 to pick you up and charges *c* cents for each mile of your trip.

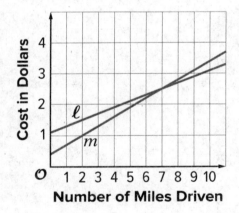

a. Match the services to the Lines ℓ and *m*.

b. For Service B, is the additional charge per mile greater or less than 30 cents per mile of the trip? Explain your reasoning.

3. Kiran and Clare like to race each other home from school. They run at the same speed, but Kiran's house is slightly closer to school than Clare's house. On a graph, their distance from their homes in meters is a function of the time from when they begin the race in seconds.

   a. As you read the graphs left to right, would the lines go up or down?

   b. What is different about the lines representing Kiran's run and Clare's run?

   c. What is the same about the lines representing Kiran's run and Clare's run?

4. Write an equation for each line. (Lesson 3-11)

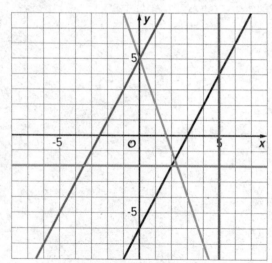

Lesson 5-9

Linear Models

NAME _____ DATE _____ PERIOD _____

Learning Goal Let's model situations with linear functions.

 ## Warm Up
9.1 Candlelight

A candle is burning. It starts out 12 inches long. After 1 hour, it is 10 inches long. After 3 hours, it is 5.5 inches long.

1. When do you think the candle will burn out completely?

2. Is the height of the candle a function of time? If yes, is it a linear function? Explain your thinking.

Activity

9.2 Shadows

When the Sun was directly overhead, the stick had no shadow. After 20 minutes, the shadow was 10.5 cm long. After 60 minutes, it was 26 cm long.

  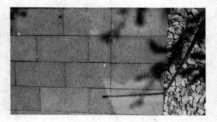

1. Based on this information, estimate how long it will be after 95 minutes.

2. After 95 minutes, the shadow measured 38.5 cm. How does this compare to your estimate?

3. Is the length of the shadow a function of time? If so, is it linear? Explain your reasoning.

NAME _____ DATE _____ PERIOD _____

Activity
9.3 Recycling

In an earlier lesson, we saw this graph that shows the percentage of all garbage in the U.S. that was recycled between 1991 and 2013.

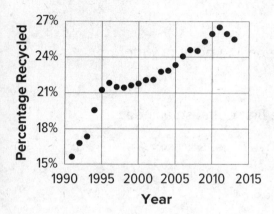

1. Sketch a linear function that models the change in the percentage of garbage that was recycled between 1991 and 1995. For which years is the model good at predicting the percentage of garbage that is produced? For which years is it not as good?

2. Pick another time period to model with a sketch of a linear function. For which years is the model good at making predictions? For which years is it not very good?

Water has different boiling points at different elevations. At 0 m above sea level, the boiling point is 100°C. At 2,500 m above sea level, the boiling point is 91.3°C. If we assume the boiling point of water is a linear function of elevation, we can use these two data points to calculate the slope of the line:

$$m = \frac{91.3 - 100}{2,500 - 0} = \frac{-8.7}{2,500}$$

This slope means that for each increase of 2,500 m, the boiling point of water decreases by 8.7°C.

Next, we already know the y-intercept is 100°C from the first point, so a linear equation representing the data is

$$y = \frac{-8.7}{2,500}x + 100$$

This equation is an example of a mathematical *model*. A mathematical model is a mathematical object like an equation, a function, or a geometric figure that we use to represent a real-life situation. Sometimes a situation can be modeled by a linear function. We have to use judgment about whether this is a reasonable thing to do based on the information we are given. We must also be aware that the model may make imprecise predictions, or may only be appropriate for certain ranges of values.

Testing our model for the boiling point of water, it accurately predicts that at an elevation of 1,000 m above sea level (when $x = 1,000$), water will boil at 96.5°C since $y = \frac{-8.7}{2,500} \cdot 1000 + 100 = 96.5$.

For higher elevations, the model is not as accurate, but it is still close. At 5,000 m above sea level, it predicts 82.6°C, which is 0.6° C off the actual value of 83.2°C. At 9,000 m above sea level, it predicts 68.7°C, which is about 3°C less than the actual value of 71.5°C.

The model continues to be less accurate at even higher elevations since the relationship between the boiling point of water and elevation isn't linear, but for the elevations in which most people live, it's pretty good.

NAME _____ DATE _____ PERIOD _____

Practice
Linear Models

1. On the first day after the new moon, 2% of the Moon's surface is illuminated. On the second day, 6% is illuminated.

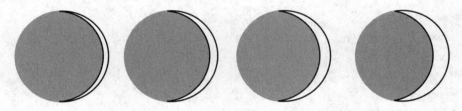

   a. Based on this information, predict the day on which the Moon's surface is 50% illuminated and 100% illuminated.

   b. The Moon's surface is 100% illuminated on day 14. Does this agree with the prediction you made?

   c. Is the percentage illumination of the Moon's surface a linear function of the day?

2. In science class, Jada uses a graduated cylinder with water in it to measure the volume of some marbles. After dropping in 4 marbles so they are all under water, the water in the cylinder is at a height of 10 milliliters. After dropping in 6 marbles so they are all under water, the water in the cylinder is at a height of 11 milliliters.

   a. What is the volume of 1 marble?

   b. How much water was in the cylinder before any marbles were dropped in?

   c. What should be the height of the water after 13 marbles are dropped in?

   d. Is the relationship between volume of water and number of marbles a linear relationship? If so, what does the slope of a line representing this relationship mean? If not, explain your reasoning.

NAME _____ DATE _____ PERIOD _____

3. Solve each of these equations. Explain or show your reasoning. **(Lesson 4-5)**

   a. $2(3x + 2) = 2x + 28$ **b.** $5y + 13 = -43 - 3y$ **c.** $4(2a + 2) = 8(2 - 3a)$

4. For a certain city, the high temperatures (in degrees Celsius) are plotted against the number of days after the new year.

   Based on this information, is the high temperature in this city a linear function of the number of days after the new year?

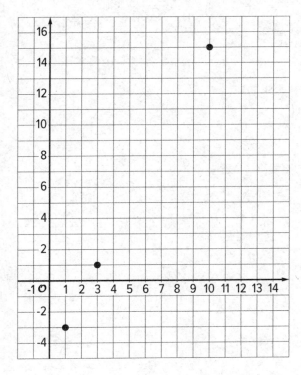

5. The school designed their vegetable garden to have a perimeter of 32 feet with the length measuring two feet more than twice the width. (Lesson 4-15)

  a. Using ℓ to represent the length of the garden and w to represent its width, write and solve a system of equations that describes this situation.

  b. What are the dimensions of the garden?

Lesson 5-10

Piecewise Linear Functions

NAME _____ DATE _____ PERIOD _____

Learning Goal Let's explore functions built out of linear pieces.

Warm Up
10.1 Notice and Wonder: Lines on Dots

What do you notice? What do you wonder?

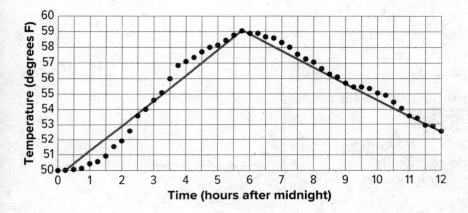

 Activity
10.2 Modeling Recycling

1. Approximate the percentage recycled each year
with a piecewise linear function by drawing
between three and five line segments to
approximate the graph.

2. Find the slope for each piece. What do these
slopes tell you?

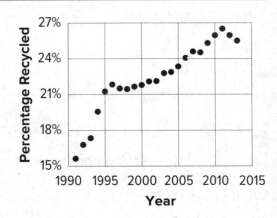

 Activity
10.3 Dog Bath

Elena filled up the tub and gave her dog a bath.
Then she let the water out of the tub.

1. The graph shows the amount of water in the
tub, in gallons, as a function of time, in
minutes. Add labels to the graph to
show this.

2. When did she turn off the water faucet?

3. How much water was in the tub when she
bathed her dog?

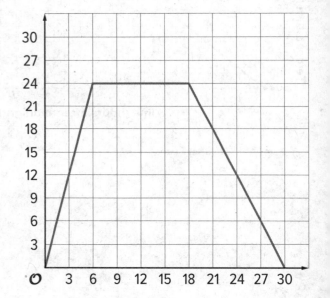

4. How long did it take for the tub to drain
completely?

5. At what rate did the faucet fill the tub?

6. At what rate did the water drain from the tub?

NAME _____ DATE _____ PERIOD _____

Activity
10.4 Distance and Speed

The graph shows the speed of a car as a function of time.

Describe what a person watching the car would see.

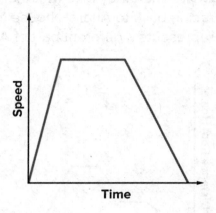

Are you ready for more?

The graph models the speed of a car over a function of time during a 3-hour trip.

How far did the car go over the course of the trip?

There is a nice way to visualize this quantity in terms of the graph. Can you find it?

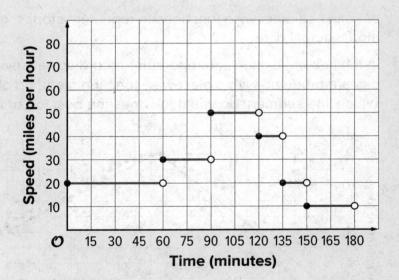

This graph shows Andre biking to his friend's house where he hangs out for a while. Then they bike together to the store to buy some groceries before racing back to Andre's house for a movie night. Each line segment in the graph represents a different part of Andre's travels.

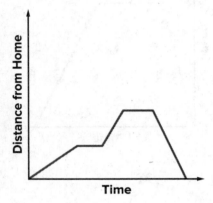

This is an example of a piecewise linear function, which is a function whose graph is pieced together out of line segments. It can be used to model situations in which a quantity changes at a constant rate for a while, then switches to a different constant rate.

We can use piecewise functions to represent stories, or we can use them to model actual data.

In the second example, temperature recordings at several times throughout a day are modeled with a piecewise function made up of two line segments. Which line segment do you think does the best job of modeling the data?

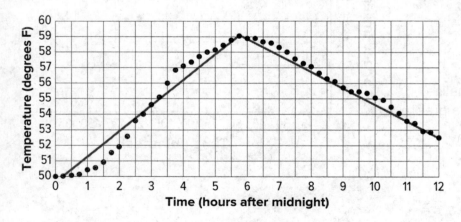

NAME _____ DATE _____ PERIOD _____

Practice
Piecewise Linear Functions

1. The graph shows the distance of a car from home as a function of time. Describe what a person watching the car may be seeing.

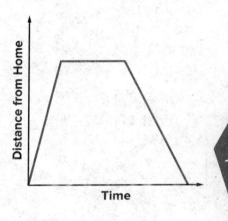

2. The equation and the graph represent two functions. Use the equation $y = 4$ and the graph to answer the questions. **(Lesson 5-7)**

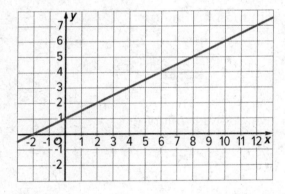

a. When x is 4, is the output of the equation or the graph greater?

b. What value for x produces the same output in both the graph and the equation?

3. This graph shows a trip on a bike trail. The trail has markers every 0.5 km showing the distance from the beginning of the trail.

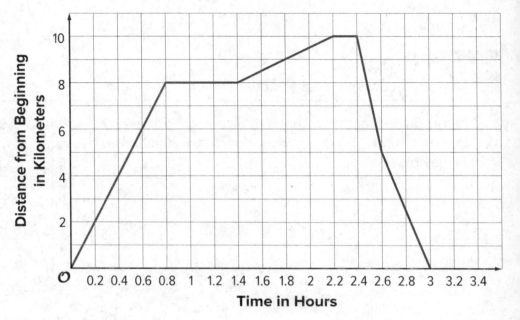

a. When was the bike rider going the fastest?

b. When was the bike rider going the slowest?

c. During what times was the rider going away from the beginning of the trail?

d. During what times was the rider going back towards the beginning of the trail?

e. During what times did the rider stop?

4. The expression $-25t + 1250$ represents the volume of liquid of a container after t seconds. The expression $50t + 250$ represents the volume of liquid of another container after t seconds. What does the equation $-25t + 1250 = 50t + 250$ mean in this situation? **(Lesson 4-9)**

Lesson 5-11

Filling Containers

NAME _____ DATE _____ PERIOD _____

Learning Goal Let's fill containers with water.

 ## Warm Up
11.1 Which One Doesn't Belong: Solids

These are drawings of three-dimensional objects. Which one doesn't belong?
Explain your reasoning.

A B C  D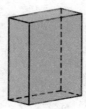

Your teacher will give you a graduated cylinder, water, and some other supplies. Your group will use these supplies to investigate the height of water in the cylinder as a function of the water volume.

1. Before you get started, make a prediction about the shape of the graph.

2 Fill the cylinder with different amounts of water and record the data in the table.

Volume (ml)						
Height (cm)						

3. Create a graph that shows the height of the water in the cylinder as a function of the water volume.

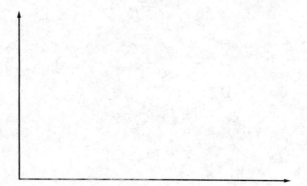

4. Choose a point on the graph and explain its meaning in the context of the situation.

NAME _____ DATE _____ PERIOD _____

Activity

11.3 What Is the Shape?

1. The graph shows the height vs. volume function of an unknown container. What shape could this container have? Explain how you know and draw a possible container.

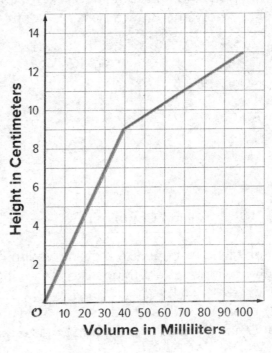

2. The graph shows the height vs. volume function of a different unknown container. What shape could this container have? Explain how you know and draw a possible container.

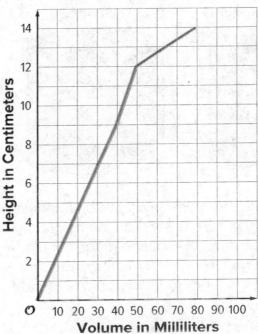

3. How are the two containers similar? How are they different?

The graph shows the height vs. volume function of an unknown container. What shape could this container have? Explain how you know and draw a possible container.

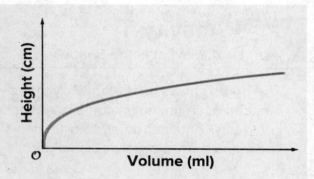

Summary
Filling Containers

When filling a shape like a cylinder with water, we can see how the dimensions of the cylinder affect things like the changing height of the water.

For example, let's say we have two cylinders, D and E, with the same height, but D has a radius of 3 cm and E has a radius of 6 cm.

Cylinder D   Cylinder E

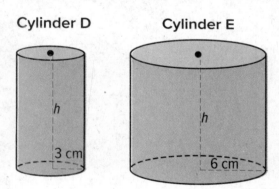

If we pour water into both cylinders at the same rate, the height of water in D will increase faster than the height of water in E due to its smaller radius. This means that if we made graphs of the height of water as a function of the volume of water for each cylinder, we would have two lines and the slope of the line for cylinder D would be greater than the slope of the line for cylinder E.

Glossary

cylinder

NAME _____ DATE _____ PERIOD _____

Practice
Filling Containers

1. Cylinder A, B, and C have the same radius but different heights. Put the cylinders in order of their volume from least to greatest.

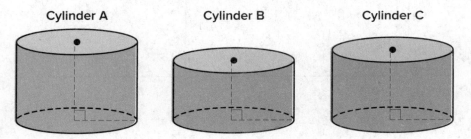

Cylinder A   Cylinder B   Cylinder C

2. Two cylinders, *a* and *b*, each started with different amounts of water. The graph shows how the height of the water changed as the volume of water increased in each cylinder. Match the graphs of *a* and *b* to Cylinders P and Q. Explain your reasoning.

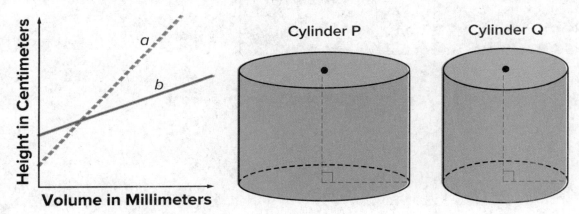

Cylinder P   Cylinder Q

3. Which of the following graphs could represent the volume of water in a cylinder as a function of its height? Explain your reasoning.

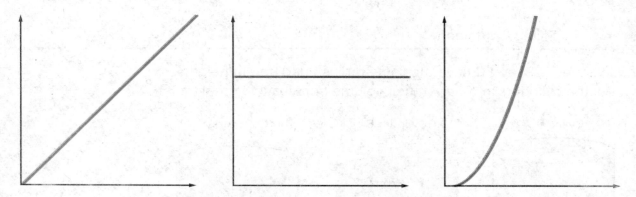

4. Together, the areas of the rectangles sum to 30 square centimeters. (Lesson 5-3)

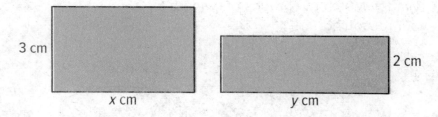

a. Write an equation showing the relationship between *x* and *y*.

b. Fill in the table with the missing values.

x	3		8		12
y		5		10	

Lesson 5-12

How Much Will Fit?

NAME _____ DATE _____ PERIOD _____

Learning Goal Let's reason about the volume of different shapes.

Warm Up
12.1 Two Containers

Your teacher will show you some containers. The small container holds 200 beans. Estimate how many beans the large jar holds.

Activity
12.2 What's Your Estimate?

Your teacher will show you some containers.

1. If the pasta box holds 8 cups of rice, how much rice would you need for the other rectangular prisms?

2. If the pumpkin can holds 15 fluid ounces of rice, how much do the other cylinders hold?

3. If the small **cone** holds 2 fluid ounces of rice, how much does the large cone hold?

4. If the golf ball were hollow, it would hold about 0.2 cups of water. If the baseball were hollow, how much would the **sphere** hold?

Activity

12.3 Do You Know These Figures?

1. What shapes are the faces of each type of object shown here? For example, all six faces of a cube are squares.

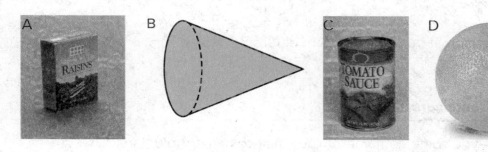

A   B   C   D

2. Which faces could be referred to as a "base" of the object?

3. Here is a method for quickly sketching a cylinder:

    •   Draw two ovals.

    •   Connect the edges.

    •   Which parts of your drawing would be hidden behind the cylinder? Make these parts dashed lines.

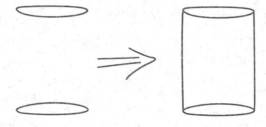

    Practice sketching some cylinders. Sketch a few different sizes, including short, tall, narrow, wide, and sideways. Label the radius r and height h on each cylinder.

Are you ready for more?

A soccer ball is a polyhedron with 12 black pentagonal faces and 20 white hexagonal faces. How many edges in total are on this polyhedron?

NAME _____ DATE _____ PERIOD _____

Summary
How Much Will Fit?

The volume of a three-dimensional figure, like a jar or a room, is the amount of space the shape encloses. We can measure volume by finding the number of equal-sized volume units that fill the figure without gaps or overlaps.
For example, we might say that a room has a volume of 1,000 cubic feet, or that a pitcher can carry 5 gallons of water. We could even measure volume of a jar by the number of beans it could hold, though a bean count is not really a measure of the volume in the same way that a cubic centimeter is because there is space between the beans. (The number of beans that fit in the jar do depend on the volume of the jar, so it is an okay estimate when judging the relative sizes of containers.)

In earlier grades, we studied three-dimensional figures with flat faces that are polygons. We learned how to calculate the volumes of rectangular prisms. Now we will study three-dimensional figures with circular faces and curved surfaces: cones, cylinders, and spheres.

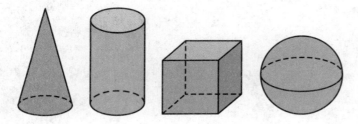

To help us see the shapes better, we can use dashed lines to represent parts that we wouldn't be able to see if a solid physical object were in front of us. For example, if we think of the cylinder in this picture as representing a tin can, the dashed arc in the bottom half of that cylinder represents the back half of the circular base of the can. What objects could the other figures in the picture represent?

Glossary

cone
sphere

Practice
How Much Will Fit?

1. a. Sketch a cube and label its side length as 4 cm (this will be Cube A).

   b. Sketch a cube with sides that are twice as long as Cube A and label its side length (this will be Cube B).

   c. Find the volumes of Cube A and Cube B.

2. Two paper drink cups are shaped like cones. The small cone can hold 6 oz of water. The large cone is $\frac{4}{3}$ the height and $\frac{4}{3}$ the diameter of the small cone. Which of these could be the amount of water the large cone holds?

   (A.) 8 cm

   (B.) 14 oz

   (C.) 4.5 oz

   (D.) 14 cm

NAME _____ DATE _____ PERIOD _____

3. The graph represents the volume of a cylinder with a height equal to its radius. **(Lesson 5-7)**

   a. When the diameter is 2 cm, what is the radius of the cylinder?

   b. Express the volume of a cube of side length *s* as an equation.

   c. Make a table for volume of the cube at $s = 0$ cm, $s = 1$ cm, $s = 2$ cm, and $s = 3$ cm.

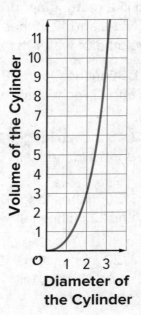

Volume of the Cylinder

Diameter of the Cylinder

   d. Which volume is greater: the volume of the cube when $s = 3$ cm, or the volume of the cylinder when its diameter is 3 cm?

4. Select **all** the points that are on a line with slope 2 that also contains the point (2, -1). **(Lesson 3-10)**

   (A.) (3, 1)

   (B.) (1, 1)

   (C.) (1, -3)

   (D.) (4, 0)

   (E.) (6, 7)

5. Several glass aquariums of various sizes are for sale at a pet shop. They are all shaped like rectangular prisms. A 15-gallon tank is 24 inches long, 12 inches wide, and 12 inches tall. Match the dimensions of the other tanks with the volume of water they can each hold.

a. Tank 1: 36 inches long, 18 inches wide, and 12 inches tall

b. Tank 2: 16 inches long, 8 inches wide, and 10 inches tall

c. Tank 3: 30 inches long, 12 inches wide, and 12 inches tall

d. Tank 4: 20 inches long, 10 inches wide, and 12 inches tall

i. 5 gallons

ii. 10 gallons

iii. 20 gallons

iv. 30 gallons

6. Solve: $\begin{cases} y = \text{-}2x - 20 \\ y = x + 4 \end{cases}$. **(Lesson 4-14)**

Lesson 5-13

The Volume of a Cylinder

NAME _____ DATE _____ PERIOD _____

Learning Goal Let's explore cylinders and their volumes.

 Warm Up

13.1 A Circle's Dimensions

Here is a circle. Points A, B, C, and D are drawn, as well as Segments AD and BC.

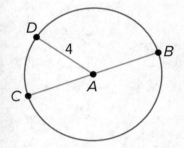

1. What is the area of the circle, in square units? Select **all** that apply.

    A. 4π

    B. $\pi 8$

    C. 16π

    D. $\pi 4^2$

    E. approximately 25

    F. approximately 50

2. If the area of a circle is 49π square units, what is its radius?
    Explain your reasoning.

Activity

13.2 Circular Volumes

What is the volume of each figure, in cubic units? Even if you aren't sure, make a reasonable guess.

Figure A   **Figure B**   **Figure C**

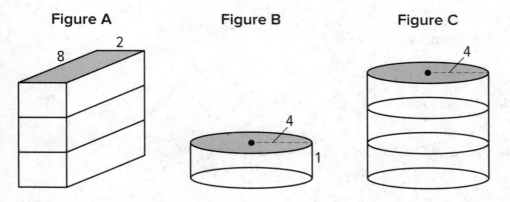

1. Figure A: A rectangular prism whose base has an area of 16 square units and whose height is 3 units.

2. Figure B: A cylinder whose base has an area of 16π square units and whose height is 1 unit.

3. Figure C: A cylinder whose base has an area of 16π square units and whose height is 3 units.

NAME _____ DATE _____ PERIOD _____

Are you ready for more?

prism	prism	prism	cylinder
base: square	base: hexagon	base: octagon	base: circle

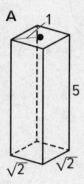
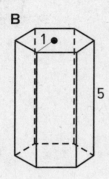
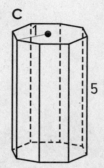
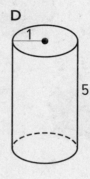

Here are solids that are related by a common measurement. In each of these solids, the distance from the center of the base to the furthest edge of the base is 1 unit, and the height of the solid is 5 units. Use 3.14 as an approximation for π to solve these problems.

1. Find the area of the square base and the circular base.

2. Use these areas to compute the volumes of the rectangular prism and the cylinder. How do they compare?

3 Without doing any calculations, list the figures from smallest to largest by volume. Use the images and your knowledge of polygons to explain your reasoning.

4. The area of the hexagon is approximately 2.6 square units, and the area of the octagon is approximately 2.83 square units. Use these areas to compute the volumes of the prisms with the hexagon and octagon bases. How does this match your explanation to the previous question?

1. For cylinders A–D, sketch a radius and the height. Label the radius with an *r* and the height with an *h*.

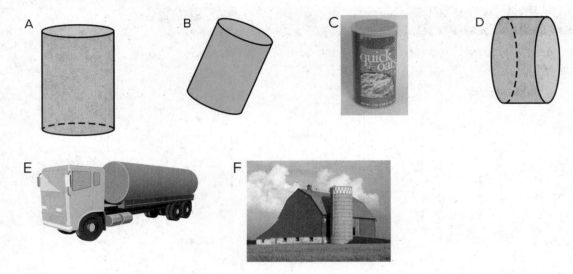

2. Earlier you learned how to sketch a cylinder. Sketch cylinders for E and F and label each one's radius and height.

NAME _____ DATE _____ PERIOD _____

Activity

13.4 A Cylinder's Volume

1. Here is a cylinder with height 4 units and diameter 10 units.

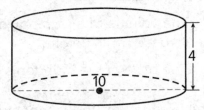

   a. Shade the cylinder's base.

   b. What is the area of the cylinder's base? Express your answer in terms of π.

   c. What is the volume of this cylinder? Express your answer in terms of π.

2. A silo is a cylindrical container that is used on farms to hold large amounts of goods, such as grain. On a particular farm, a silo has a height of 18 feet and diameter of 6 feet. Make a sketch of this silo and label its height and radius. How many cubic feet of grain can this silo hold? Use 3.14 as an approximation for π.

Are you ready for more?

One way to construct a cylinder is to take a rectangle (for example, a piece of paper), curl two opposite edges together, and glue them in place.

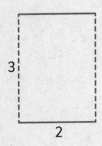

Which would give the cylinder with the greater volume: Gluing the two dashed edges together, or gluing the two solid edges together?

We can find the volume of a cylinder with radius r and height h using two ideas we've seen before:

- The volume of a rectangular prism is a result of multiplying the area of its base by its height.

- The base of the cylinder is a circle with radius r, so the base area is πr^2.

Remember that π is the number we get when we divide the circumference of any circle by its diameter. The value of π is approximately 3.14.

Just like a rectangular prism, the volume of a cylinder is the area of the base times the height.

For example, take a cylinder whose radius is 2 cm and whose height is 5 cm.

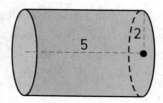

The base has an area of 4π cm^2 (since $\pi \cdot 2^2 = 4\pi$), so the volume is 20π cm^3 (since $4\pi \cdot 5 = 20\pi$). Using 3.14 as an approximation for π, we can say that the volume of the cylinder is approximately 62.8 cm^3.

In general, the base of a cylinder with radius r units has area πr^2 square units. If the height is h units, then the volume V in cubic units is:

$$V = \pi r^2 h$$

Practice
The Volume of a Cylinder

1. Respond to each of the following.
   a. Sketch a cylinder.

   b. Label its radius 3 and its height 10.

   c. Shade in one of its bases.

2. At a farm, animals are fed bales of hay and buckets of grain. Each bale of hay is in the shape of a rectangular prism. The base has side lengths 2 feet and 3 feet, and the height is 5 feet. Each bucket of grain is a cylinder with a diameter of 3 feet. The height of the bucket is 5 feet, the same as the height of the bale.

   a. Which is larger in area, the rectangular base of the bale or the circular base of the bucket? Explain how you know.

   b. Which is larger in volume, the bale or the bucket? Explain how you know.

3. Three cylinders have a height of 8 cm. Cylinder 1 has a radius of 1 cm. Cylinder 2 has a radius of 2 cm. Cylinder 3 has a radius of 3 cm. Find the volume of each cylinder.

4. A one-quart container of tomato soup is shaped like a rectangular prism. A soup bowl shaped like a hemisphere can hold 8 oz of liquid. How many bowls will the soup container fill? Recall that 1 quart is equivalent to 32 fluid ounces (oz). **(Lesson 5-12)**

5. Match each set of information about a circle with the area of that circle.

   a. Circle A has a radius of 4 units.   i. 4π square units

   b. Circle B has a radius of 10 units.   ii. approximately 314 square units

   c. Circle C has a diameter of 16 units.   iii. 64π square units

   d. Circle D has a circumference of 4π units.   iv. 16π square units

6. Two students join a puzzle solving club and get faster at finishing the puzzles as they get more practice. Student A improves their times faster than Student B. **(Lesson 5-8)**

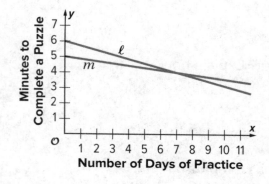

   a. Match the students to the Lines ℓ and m.

   b. Which student was faster at puzzle solving before practice?

Lesson 5-14

Finding Cylinder Dimensions

NAME _____ DATE _____ PERIOD _____

Learning Goal Let's figure out the dimensions of cylinders.

 ## Warm Up
14.1 A Cylinder of Unknown Height

What is a possible volume for this cylinder if the diameter is 8 cm? Explain your reasoning.

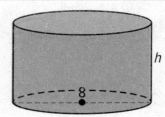

 ## Activity
14.2 What's the Dimension?

The volume V of a cylinder with radius r is given by the formula $V = \pi r^2 h$.

1. The volume of this cylinder with radius 5 units is 50π cubic units. This statement is true: $50\pi = 5^2\pi h$.

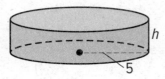

   What does the height of this cylinder have to be? Explain how you know.

2. The volume of this cylinder with height 4 units is 36π cubic units. This statement is true: $36\pi = r^2\pi4$.

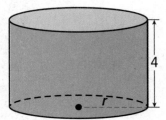

What does the radius of this cylinder have to be? Explain how you know.

Are you ready for more?

Suppose a cylinder has a volume of 36π cubic inches, but it is not the same cylinder as the one you found earlier in this activity.

1. What are some possibilities for the dimensions of the cylinder?

2. How many different cylinders can you find that have a volume of 36π cubic inches?

NAME _____ DATE _____ PERIOD _____

Activity
14.3 Cylinders with Unknown Dimensions

Each row of the table has information about a particular cylinder. Complete the
table with the missing dimensions.

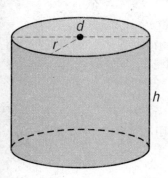

Diameter (units)	Radius (units)	Area of the Base (square units)	Height (units)	Volume (cubic units)
	3		5	
12				108π
			11	99π
8				16π
			100	16π
	10			20π
20				314
			b	$\pi \cdot b \cdot a^2$

In an earlier lesson we learned that the volume, V, of a cylinder with radius r and height h is ...

$$V = \pi r^2 h$$

We say that the volume depends on the radius and height, and if we know the radius and height, we can find the volume. It is also true that if we know the volume and one dimension (either radius or height), we can find the other dimension.

For example, imagine a cylinder that has a volume of 500π cm^3 and a radius of 5 cm, but the height is unknown. From the volume formula we know that. . .

$$500\pi = \pi \cdot 25 \cdot h$$

must be true. Looking at the structure of the equation, we can see that $500 = 25h$. That means that the height has to be 20 cm, since $500 \div 25 = 20$.

Now imagine another cylinder that also has a volume of 500π cm^3 with an unknown radius and a height of 5 cm. Then we know that. . .

$$500\pi = \pi \cdot r^2 \cdot 5$$

must be true. Looking at the structure of this equation, we can see that $r^2 = 100$. So the radius must be 10 cm.

NAME _____ DATE _____ PERIOD _____

Practice
Finding Cylinder Dimensions

1. Complete the table with all of the missing information about three different cylinders.

Diameter of Base (units)	Area of Base (square units)	Height (units)	Volume (cubic units)
4		10	
6			63π
	25π	6	

2. A cylinder has volume 45π and radius 3. What is its height?

3. Three cylinders have a volume of 2826 cm^3. Cylinder A has a height of 900 cm. Cylinder B has a height of 225 cm. Cylinder C has a height of 100 cm. Find the radius of each cylinder. Use 3.14 as an approximation for π.

4. A gas company's delivery truck has a cylindrical tank that is 14 feet in diameter and 40 feet long. (Lesson 5-13)

   a. Sketch the tank, and mark the radius and the height.

   b. How much gas can fit in the tank?

5. Here is a graph that shows the water height of the ocean between September 22 and September 24, 2016 in Bodega Bay, CA. (Lesson 5-5)

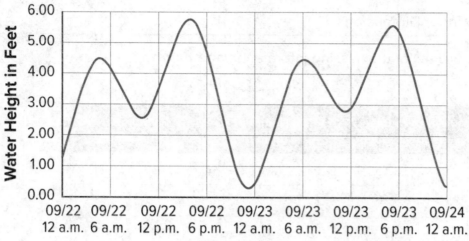

a. Estimate the water height at 12 p.m. on September 22.

b. How many times was the water height 5 feet? Find two times when this happens.

c. What was the lowest the water got during this time period? When does this occur?

d. Does the water ever reach a height of 6 feet?

Lesson 5-15

The Volume of a Cone

NAME _____ DATE _____ PERIOD _____

Learning Goal Let's explore cones and their volumes.

Warm Up
15.1 Which Has a Larger Volume?

The cone and cylinder have the same height, and the radii of their bases are equal.

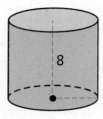 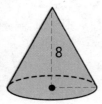

1. Which figure has a larger volume?

2. Do you think the volume of the smaller one is more or less than $\frac{1}{2}$ the volume of the larger one? Explain your reasoning.

3. Sketch two different sized cones. The oval doesn't have to be on the bottom! For each drawing, label the cone's radius with *r* and height with *h*.

Here is a method for quickly sketching a cone:

- Draw an oval.
- Draw a point centered above the oval.
- Connect the edges of the oval to the point.
- Which parts of your drawing would be hidden behind the object? Make these parts dashed lines.

Activity

15.2 From Cylinders to Cones

A cone and cylinder have the same height and their bases are congruent circles.

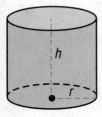 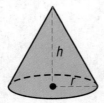

1. If the volume of the cylinder is 90 cm³, what is the volume of the cone?

2. If the volume of the cone is 120 cm³, what is the volume of the cylinder?

3. If the volume of the cylinder is $V = \pi r^2 h$, what is the volume of the cone? Either write an expression for the cone or explain the relationship in words.

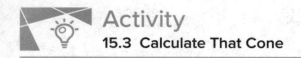
1. Here is a cylinder and cone that have the same height and the same base area. What is the volume of each figure? Express your answers in terms of π.

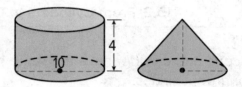

2. Here is a cone.

    a. What is the area of the base? Express your answer in terms of π.

    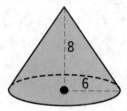

    b. What is the volume of the cone? Express your answer in terms of π.

3. A cone-shaped popcorn cup has a radius of 5 centimeters and a height of 9 centimeters. How many cubic centimeters of popcorn can the cup hold? Use 3.14 as an approximation for π, and give a numerical answer.

NAME _____ DATE _____ PERIOD _____

Are you ready for more?

A grain silo has a cone shaped spout on the bottom in order to regulate the flow of grain out of the silo. The diameter of the silo is 8 feet. The height of the cylindrical part of the silo above the cone spout is 12 feet while the height of the entire silo is 16 feet.

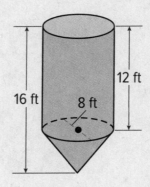

How many cubic feet of grain are held in the cone spout of the silo? How many cubic feet of grain can the entire silo hold?

 ## Summary
The Volume of a Cone

If a cone and a cylinder have the same base and the same height, then the volume of the cone is $\frac{1}{3}$ of the volume of the cylinder.

For example, the cylinder and cone shown here both have a base with radius 3 feet and a height of 7 feet.

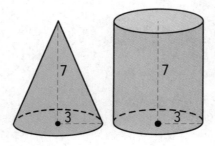

The cylinder has a volume of 63π cubic feet since $\pi \cdot 3^2 \cdot 7 = 63\pi$.
The cone has a volume that is $\frac{1}{3}$ of that, or 21π cubic feet.

If the radius for both is r and the height for both is h, then the volume of the cylinder is $\pi r^2 h$. That means that the volume, V, of the cone is. . .

$$V = \frac{1}{3}\pi r^2 h$$

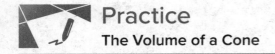

Practice
The Volume of a Cone

1. A cylinder and cone have the same height and radius. The height of each is 5 cm, and the radius is 2 cm. Calculate the volume of the cylinder and the cone.

2. The volume of this cone is 36π cubic units. What is the volume of a cylinder that has the same base area and the same height?

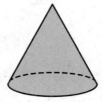

3. A cylinder has a diameter of 6 cm and a volume of 36π cm^3. **(Lesson 5-14)**

   a. Sketch the cylinder.

   b. Find its height and radius in centimeters.

   c. Label your sketch with the cylinder's height and radius.

4. Lin wants to get some custom T-shirts printed for her basketball team. Shirts cost $10 each if you order 10 or fewer shirts and $9 each if you order 11 or more shirts. (Lesson 5-10)

a. Make a graph that shows the total cost of buying shirts, for 0 through 15 shirts.

b. There are 10 people on the team. Do they save money if they buy an extra shirt? Explain your reasoning.

c. What is the slope of the graph between 0 and 10? What does it mean in the story?

d. What is the slope of the graph between 11 and 15? What does it mean in the story?

5. In the following graphs, the horizontal axis represents time and the vertical axis represents distance from school. Write a possible story for each graph. (Lesson 5-6)

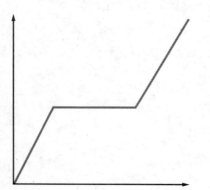 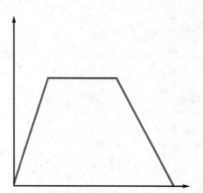 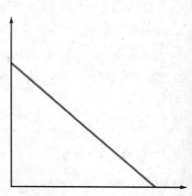

Lesson 5-16

Finding Cone Dimensions

NAME _____ DATE _____ PERIOD _____

Learning Goal Let's figure out the dimensions of cones.

Warm Up
16.1 Number Talk: Thirds

For each equation, decide what value, if any, would make it true.

1. $27 = \frac{1}{3}h$

2. $27 = \frac{1}{3}r^2$

3. $12\pi = \frac{1}{3}\pi a$

4. $12\pi = \frac{1}{3}\pi b^2$

Activity
16.2 An Unknown Radius

The volume V of a cone with radius r is given by the formula $V = \frac{1}{3}\pi r^2 h$.

The volume of this cone with height 3 units and radius r is $V = 64\pi$ cubic units.

This statement is true: $64\pi = \frac{1}{3}\pi r^2 \cdot 3$.

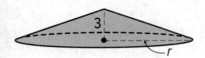

What does the radius of this cone have to be? Explain how you know.

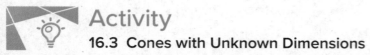

Activity

16.3 Cones with Unknown Dimensions

Each row of the table has some information about a particular cone. Complete the table with the missing dimensions.

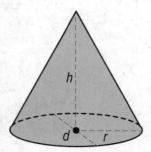

Diameter (units)	Radius (units)	Area of the Base (square units)	Height (units)	Volume of Cone (cubic units)
	4		3	
	$\frac{1}{3}$		6	
		144π	$\frac{1}{4}$	
20				200π
			12	64π
			3	3.14

NAME _____ DATE _____ PERIOD _____

A *frustum* is the result of taking a cone and slicing off a smaller cone using a cut parallel to the base.

Find a formula for the volume of a frustum, including deciding which quantities you are going to include in your formula.

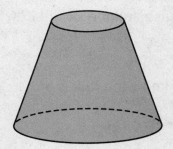

Activity
16.4 Popcorn Deals

A movie theater offers two containers:

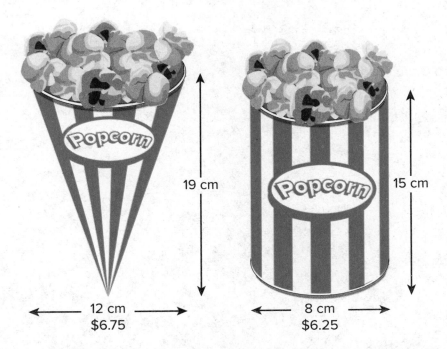

19 cm

15 cm

12 cm
$6.75

8 cm
$6.25

Which container is the better value? Use 3.14 as an approximation for π.

As we saw with cylinders, the volume V of a cone depends on the radius r of the base and the height h:

$$V = \frac{1}{3}\pi r^2 h$$

If we know the radius and height, we can find the volume. If we know the volume and one of the dimensions (either radius or height), we can find the other dimension.

For example, imagine a cone with a volume of 64π cm^3, a height of 3 cm, and an unknown radius r. From the volume formula, we know that. . .

$$64\pi = \frac{1}{3}\pi r^2 \cdot 3$$

Looking at the structure of the equation, we can see that $r^2 = 64$, so the radius must be 8 cm.

Now imagine a different cone with a volume of 18π cm^3, a radius of 3 cm, and an unknown height h. Using the formula for the volume of the cone, we know that. . .

$$18\pi = \frac{1}{3}\pi 3^2 h$$

so the height must be 6 cm. Can you see why?

NAME _____ DATE _____ PERIOD _____

Practice
Finding Cone Dimensions

1. The volume of this cylinder is 175π cubic units. What is the volume of a cone that has the same base area and the same height? **(Lesson 5-15)**

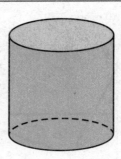

2. A cone has volume 12π cubic inches. Its height is 4 inches. What is its radius?

3. A cone has volume 3π.

   a. If the cone's radius is 1, what is its height?

   b. If the cone's radius is 2, what is its height?

   c. If the cone's radius is 5, what is its height?

   d. If the cone's radius is $\frac{1}{2}$, what is its height?

   e. If the cone's radius is r, then what is the height?

4. Three people are playing near the water. Person A stands on the dock. Person B starts at the top of a pole and ziplines into the water, then climbs out of the water. Person C climbs out of the water and up the zipline pole. Match the people to the graphs where the horizontal axis represents time in seconds and the vertical axis represents height above the water level in feet. (Lesson 5-6)

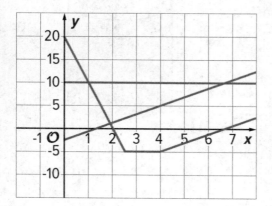

5. A room is 15 feet tall. An architect wants to include a window that is 6 feet tall. The distance between the floor and the bottom of the window is b feet. The distance between the ceiling and the top of the window is a feet. This relationship can be described by the equation $a = 15 - (b + 6)$.
(Lesson 5-3)

    a. Which variable is independent based on the equation given?

    b. If the architect wants b to be 3, what does this mean? What value of a would work with the given value for b?

    c. The customer wants the window to have 5 feet of space above it. Is the customer describing a or b? What is the value of the other variable?

Lesson 5-17

Scaling One Dimension

NAME _____ DATE _____ PERIOD _____

Learning Goal Let's see how changing one dimension changes the volume of a shape.

Warm Up

17.1 Driving the Distance

Here is a graph of the amount of gas burned during a trip by a tractor-trailer truck as it drives at a constant speed down a highway:

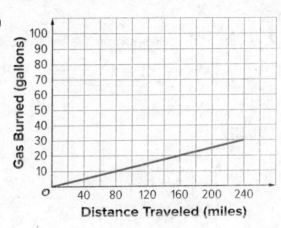

1. At the end of the trip, how far had the truck driven, and how much gas did it use?

2. If a truck traveled half this distance at the same rate, how much gas would it use?

3. If a truck traveled double this distance at the same rate, how much gas would it use?

4. Complete the sentence:
   _____ is a function of _____.

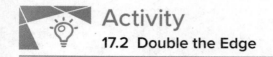
There are many right rectangular prisms with one edge of length 5 units and another edge of length 3 units. Let s represent the length of the third edge and V represent the volume of these prisms.

1. Write an equation that represents the relationship between V and s.

2. Graph this equation and label the axes.

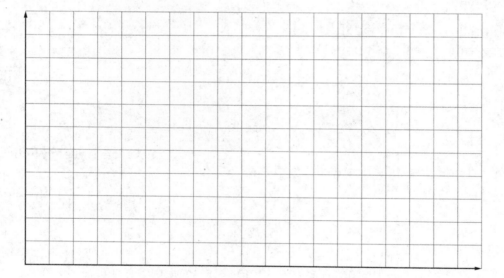

3. What happens to the volume if you double the edge length s?
   Where do you see this in the graph? Where do you see it algebraically?

NAME _____ DATE _____ PERIOD _____

Activity

17.3 Halve the Height

There are many cylinders with radius 5 units. Let *h* represent the height and *V* represent the volume of these cylinders.

1. Write an equation that represents the relationship between *V* and *h*. Use 3.14 as an approximation of π.

2. Graph this equation and label the axes.

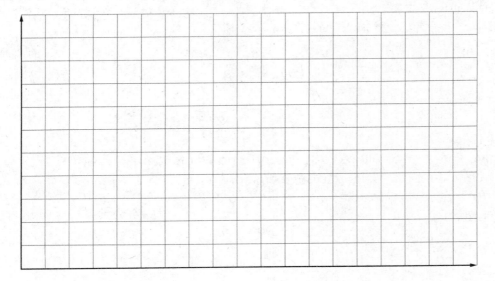

3. What happens to the volume if you halve the height, *h*? Where can you see this in the graph? How can you see it algebraically?

Suppose we have a rectangular prism with dimensions 2 units by 3 units by 6 units, and we would like to make a rectangular prism of volume 216 cubic units by stretching *one* of the three dimensions.

- What are the three ways of doing this? Of these, which gives the prism with the smallest surface area?

- Repeat this process for a starting rectangular prism with dimensions 2 units by 6 units by 6 units.

- Can you give some general tips to someone who wants to make a box with a certain volume, but wants to save cost on material by having as small a surface area as possible?

NAME _____ DATE _____ PERIOD _____

Activity
17.4 Figuring Out Cone Dimensions

Here is a graph of the relationship between the height and the volume of some cones that all have the same radius:

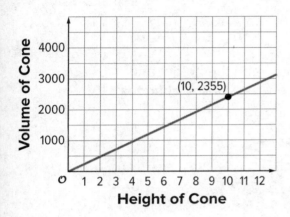

1. What do the coordinates of the labeled point represent?

2. What is the volume of the cone with height 5? With height 30?

3. Use the labeled point to find the radius of these cones. Use 3.14 as an approximation for π.

4. Write an equation that relates the volume V and height h.

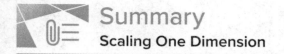

Imagine a cylinder with a radius of 5 cm that is being filled with water. As the height of the water increases, the volume of water increases.

We say that the volume of the water in the cylinder, V, depends on the height of the water, h. We can represent this relationship with an equation: $V = \pi \cdot 5^2 h$ or just $V = 25\pi h$.

This equation represents a *proportional relationship* between the height and the volume. We can use this equation to understand how the volume changes when the height is tripled.

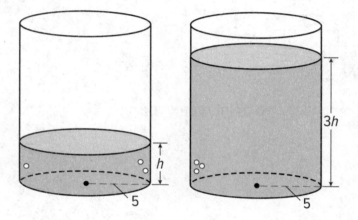

The new volume would be $V = 25\pi(3h) = 75\pi h$, which is precisely 3 times as much as the old volume of $25\pi h$. In general, when one quantity in a proportional relationship changes by a given factor, the other quantity changes by the same factor.

Remember that proportional relationships are examples of linear relationships, which can also be thought of as functions. So in this example V, the volume of water in the cylinder, is a function of the height h of the water.

NAME _____ DATE _____ PERIOD _____

Practice
Scaling One Dimension

1. A cylinder has a volume of 48π cm^3 and height h. Complete this table for volume of cylinders with the same radius but different heights.

Height (cm)	Volume (cm^3)
h	48π
$2h$	
$5h$	
$\dfrac{h}{2}$	
$\dfrac{h}{5}$	

2. A cylinder has a radius of 3 cm and a height of 5 cm.

   a. What is the volume of the cylinder?

   b. What is the volume of the cylinder when its height is tripled?

   c. What is the volume of the cylinder when its height is halved?

3. A graduated cylinder that is 24 cm tall can hold 1 L of water. What is the radius of the cylinder? What is the height of the 500 ml mark? The 250 ml mark? Recall that 1 liter (L) is equal to 1,000 milliliters (ml) and that 1 liter (L) is equal to 1,000 cm^3.

4. An ice cream shop offers two ice cream cones. The waffle cone holds 12 ounces and is 5 inches tall. The sugar cone also holds 12 ounces and is 8 inches tall. Which cone has a larger radius? (Lesson 5-16)

5. A 6 oz paper cup is shaped like a cone with a diameter of 4 inches. How many ounces of water will a plastic cylindrical cup with a diameter of 4 inches hold if it is the same height as the paper cup? (Lesson 5-15)

6. Lin's smart phone was fully charged when she started school at 8:00 a.m. At 9:20 a.m., it was 90% charged, and at noon, it was 72% charged. (Lesson 5-9)

   a. When do you think her battery will die?

   b. Is battery life a function of time? If yes, is it a linear function? Explain your reasoning.

Lesson 5-18

Scaling Two Dimensions

NAME _____ DATE _____ PERIOD _____

Learning Goal Let's change more dimensions of shapes.

 ## Warm Up
18.1 Tripling Statements

m, n, a, b, and c all represent positive integers. Consider these two equations:

$$m = a + b + c$$
$$n = abc$$

1. Which of these statements are true? Select **all** that apply.

   (A.) If a is tripled, m is tripled.

   (B.) If a, b, and c are all tripled, then m is tripled.

   (C.) If a is tripled, n is tripled.

   (D.) If a, b, and c are all tripled, then n is tripled.

2. Create a true statement of your own about one of the equations.

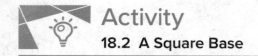

Activity

18.2 A Square Base

Clare sketches a rectangular prism with a height of 11 and a square base and labels the edges of the base *s*. She asks Han what he thinks will happen to the volume of the rectangular prism if she triples *s*.

Han says the volume will be 9 times bigger. Is he right? Explain or show your reasoning.

Are you ready for more?

A cylinder can be constructed from a piece of paper by curling it so that you can glue together two opposite edges (the dashed edges in the figure).

1. If you wanted to increase the volume inside the resulting cylinder, would it make more sense to double *x*, *y*, or does it not matter?

2. If you wanted to increase the surface area of the resulting cylinder, would it make more sense to double *x*, *y*, or does it not matter?

3. How would your answers to these questions change if we made a cylinder by gluing together the solid lines instead of the dashed lines?

NAME _____ DATE _____ PERIOD _____

Activity

18.3 Playing with Cones

There are many cones with a height of 7 units. Let r represent the radius and V represent the volume of these cones.

1. Write an equation that expresses the relationship between V and r. Use 3.14 as an approximation for π.

2. Predict what happens to the volume if you triple the value of r.

3. Graph this equation.

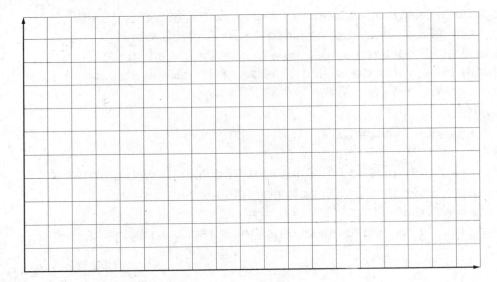

4. What happens to the volume if you triple r? Where do you see this in the graph? How can you see it algebraically?

There are many rectangular prisms that have a length of 4 units and width of 5 units but differing heights. If h represents the height, then the volume V of such a prism is ...

$$V = 20h$$

The equation shows us that the volume of a prism with a base area of 20 square units is a linear function of the height. Because this is a proportional relationship, if the height gets multiplied by a factor of a, then the volume is also multiplied by a factor of a:

$$V = 20(ah)$$

What happens if we scale *two* dimensions of a prism by a factor of a? In this case, the volume gets multiplied by a factor of a twice, or a^2.

For example, think about a prism with a length of 4 units, width of 5 units, and height of 6 units. Its volume is 120 cubic units since $4 \cdot 5 \cdot 6 = 120$. Now imagine the length and width each get scaled by a factor of a, meaning the new prism has a length of $4a$, width of $5a$, and a height of 6. The new volume is $120a^2$ cubic units since $4a \cdot 5a \cdot 6 = 120a^2$.

A similar relationship holds for cylinders.

Think of a cylinder with a height of 6 and a radius of 5. The volume would be 150π cubic units since $\pi \cdot 5^2 \cdot 6 = 150\pi$. Now, imagine the radius is scaled by a factor of a. Then the new volume is $\pi \cdot (5a)^2 \cdot 6 = \pi \cdot 25a^2 \cdot 6$ or $150a^2\pi$ cubic units. So scaling the radius by a factor of a has the effect of multiplying the volume by a^2!

Why does the volume multiply by a^2 when only the radius changes?

This makes sense if we imagine how scaling the radius changes the base area of the cylinder. As the radius increases, the base area gets larger in two dimensions (the circle gets wider and also taller), while the third dimension of the cylinder, height, stays the same.

NAME _____ DATE _____ PERIOD _____

Practice
Scaling Two Dimensions

1. There are many cylinders with a height of 18 meters. Let r represent the radius in meters and V represent the volume in cubic meters.

   a. Write an equation that represents the volume V as a function of the radius r.

   b. Complete this table, giving three possible examples.

r	V
1	

   c. If the radius of a cylinder is doubled, does the volume double? Explain how you know.

   d. Is the graph of this function a line? Explain how you know.

2. As part of a competition, Diego must spin around in a circle 6 times and then run to a tree. The time he spends on each spin is represented by s and the time he spends running is r. He gets to the tree 21 seconds after he starts spinning. **(Lesson 5-3)**

   a. Write an equation showing the relationship between s and r.

   b. Rearrange the equation so that it shows r as a function of s.

   c. If it takes Diego 1.2 seconds to spin around each time, how many seconds did he spend running?

3. The table and graph represent two functions. Use the table and graph to answer the questions. (Lesson 5-7)

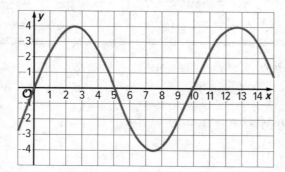

x	1	2	3	4	5	6
y	3	-1	0	4	5	-1

a. For which values of x is the output from the table less than the output from the graph?

b. In the graphed function, which values of x give an output of 0?

4. A cone has a radius of 3 units and a height of 4 units.

a. What is this volume of this cone?

b. Another cone has quadruple the radius, and the same height. How many times larger is the new cone's volume?

Lesson 5-19

Estimating a Hemisphere

NAME _____ DATE _____ PERIOD _____

Learning Goal Let's estimate volume of hemispheres with figures we know.

Warm Up
19.1 Notice and Wonder: Two Shapes

Here are two shapes.

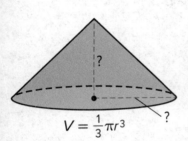

$V = \frac{1}{3}\pi r^3$

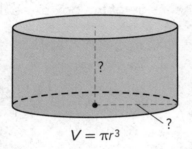

$V = \pi r^3$

What do you notice? What do you wonder?

1. Mai has a dome paperweight that she can use as a magnifier. The paperweight is shaped like a hemisphere made of solid glass, so she wants to design a box to keep it in so it won't get broken. Her paperweight has a radius of 3 cm.

   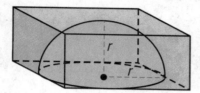

   a. What should the dimensions of the inside of the box be so the box is as small as possible?

   b. What is the volume of the box?

   c. What is a reasonable estimate for the volume of the paperweight?

2. Tyler has a different box with side lengths that are twice as long as the sides of Mai's box. Tyler's box is just large enough to hold a different glass paperweight.

   a. What is the volume of the new box?

   b. What is a reasonable estimate for the volume of this glass paperweight?

   c. How many times bigger do you think the volume of the paperweight in this box is than the volume of Mai's paperweight? Explain your thinking.

NAME _____ DATE _____ PERIOD _____

Activity
19.3 Estimating Hemispheres

1. A hemisphere with radius 5 units fits snugly into a cylinder of the same radius and height.

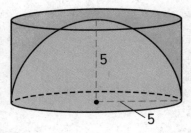

a. Calculate the volume of the cylinder.

b. Estimate the volume of the hemisphere. Explain your reasoning.

2. A cone fits snugly inside a hemisphere, and they share a radius of 5.

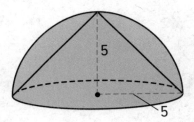

a. What is the volume of the cone?

b. Estimate the volume of the hemisphere. Explain your reasoning.

3. Compare your estimate for the hemisphere with the cone inside to your estimate of the hemisphere inside the cylinder. How do they compare to the volumes of the cylinder and the cone?

Are you ready for more?

Estimate what fraction of the volume of the cube is occupied by the pyramid that shares the base and a top vertex with the cube, as in the figure.

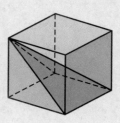

NAME _____ DATE _____ PERIOD _____

Summary
Estimating a Hemisphere

We can estimate the volume of a hemisphere by comparing it to other shapes for which we know the volume.

For example, a hemisphere of radius 1 unit fits inside a cylinder with a radius of 1 unit and height of 1 unit.

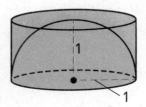

Since the hemisphere is *inside* the cylinder, it must have a smaller volume than the cylinder making the cylinder's volume a reasonable over-estimate for the volume of the hemisphere.

The volume of this particular cylinder is about 3.14 units^3 since $\pi(1)^2(1) = \pi$, so we know the volume of the hemisphere is less than 3.14 cubic units.

Using similar logic, a cone of radius 1 unit and height 1 unit fits inside of the hemisphere of radius 1 unit.

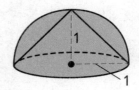

Since the cone is *inside* the hemisphere, the cone must have a smaller volume than the hemisphere making the cone's volume a reasonable under-estimate for the volume of the hemisphere.

The volume of this particular cone is about 1.05 units^3 since $\frac{1}{3}\pi(1)^2(1) = \frac{1}{3}\pi \approx 1.05$, so we know the volume of the hemisphere is more than 1.05 cubic units.

Averaging the volumes of the cylinder and the cone, we can estimate the volume of the hemisphere to be about 2.10 units^3 since $\frac{3.14 + 1.05}{2} \approx 2.10$.

And, since a hemisphere is half of a sphere, we can also estimate that a sphere with radius of 1 would be double this volume, or about 4.20 units^3.

Practice
Estimating a Hemisphere

1. A baseball fits snugly inside a transparent display cube. The length of an edge of the cube is 2.9 inches.

   Is the baseball's volume greater than, less than, or equal to 2.9^3 cubic inches? Explain how you know.

2. There are many possible cones with a height of 18 meters. Let r represent the radius in meters and V represent the volume in cubic meters.
   (Lesson 5-18)

   a. Write an equation that represents the volume V as a function of the radius r.

   b. Complete this table for the function, giving three possible examples.

r	V
2	

   c. If you double the radius of a cone, does the volume double? Explain how you know.

   d. Is the graph of this function a line? Explain how you know.

NAME _____ DATE _____ PERIOD _____

3. A hemisphere fits snugly inside a cylinder with a radius of 6 cm.
A cone fits snugly inside the same hemisphere.

   a. What is the volume of the cylinder?

   b. What is the volume of the cone?

   c. Estimate the volume of the hemisphere by calculating the average
of the volumes of the cylinder and cone.

4. a. Find the hemisphere's diameter if its radius is 6 cm.

   b. Find the hemisphere's diameter if its radius is $\frac{1,000}{3}$ m.

   c. Find the hemisphere's diameter if its radius is 9.008 ft.

   d. Find the hemisphere's radius if its diameter is 6 cm.

   e. Find the hemisphere's radius if its diameter is $\frac{1,000}{3}$ m.

   f. Find the hemisphere's radius if its diameter is 9.008 ft.

5. After almost running out of space on her phone, Elena checks with a couple of friends who have the same phone to see how many pictures they have on their phones and how much memory they take up. The results are shown in the table. (Lesson 5-9)

Number of Photos	2,523	3,148	1,875
Memory Used in MB	8,072	10,106	6,037

a. Could this information be reasonably modeled with a linear function? Explain your reasoning.

b. Elena needs to delete photos to create 1,200 MB of space. Estimate the number of photos should she delete.

Lesson 5-20

The Volume of a Sphere

NAME _____ DATE _____ PERIOD _____

Learning Goal Let's explore spheres and their volumes.

Warm Up
20.1 Sketch a Sphere

Here is a method for quickly sketching a sphere:

- Draw a circle.

- Draw an oval in the middle whose edges touch the sphere.

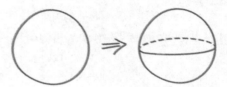

1. Practice sketching some spheres. Sketch a few different sizes.

2. For each sketch, draw a radius and label it *r*.

Activity

20.2 A Sphere in a Cylinder

Here are a cone, a sphere, and a cylinder that all have the same radii and heights. The radius of the cylinder is 5 units. When necessary, express all answers in terms of π.

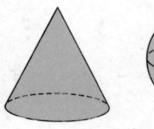 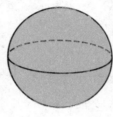 

1. What is the height of the cylinder?

2. What is the volume of the cylinder?

3. What is the volume of the cone?

4. What is the volume of the sphere? Explain your reasoning.

NAME _____ DATE _____ PERIOD _____

Activity
20.3 Spheres in Cylinders

Here are a cone, a sphere, and a cylinder that all have the same radii and heights. Let the radius of the cylinder be r units. When necessary, express answers in terms of π.

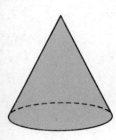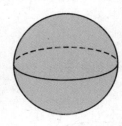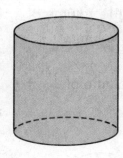

1. What is the height of the cylinder in terms of r?

2. What is the volume of the cylinder in terms of r?

3. What is the volume of the cone in terms of r?

4. What is the volume of the sphere in terms of r?

5. A volume of the cone is $\frac{1}{3}$ the volume of a cylinder. The volume of the sphere is what fraction of the volume of the cylinder?

Summary

The Volume of a Sphere

Think about a sphere with radius r units that fits snugly inside a cylinder. The cylinder must then also have a radius of r units and a height of $2r$ units. Using what we have learned about volume, the cylinder has a volume of $\pi r^2 h = \pi r^2 \cdot (2r)$, which is equal to $2\pi r^3$ cubic units.

We know from an earlier lesson that the volume of a cone with the same base and height as a cylinder has $\frac{1}{3}$ of the volume.

In this example, such a cone has a volume of $\frac{1}{3} \cdot \pi r^2 \cdot 2r$ or just $\frac{2}{3}\pi r^3$ cubic units.

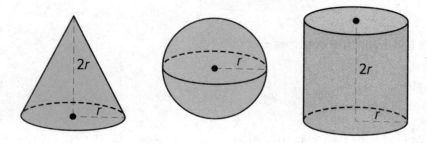

If we filled the cone and sphere with water, and then poured that water into the cylinder, the cylinder would be completely filled. That means the volume of the sphere and the volume of the cone add up to the volume of the cylinder. In other words, if V is the volume of the sphere, then. . .

$$V + \frac{2}{3}\pi r^3 = 2\pi r^3$$

This leads to the formula for the volume of the sphere:

$$V = \frac{4}{3}\pi r^3$$

NAME _____ DATE _____ PERIOD _____

Practice
The Volume of a Sphere

1. a. A cube's volume is 512 cubic units. What is the length of its edge?

b. If a sphere fits snugly inside this cube, what is its volume?

c. What fraction of the cube is taken up by the sphere? What percentage is this? Explain or show your reasoning.

2. Sphere A has radius 2 cm. Sphere B has radius 4 cm.

a. Calculate the volume of each sphere.

b. The radius of Sphere B is double that of Sphere A. How many times greater is the volume of B?

3. Three cones have a volume of 192π cm^3. Cone A has a radius of 2 cm. Cone B has a radius of 3 cm. Cone C has a radius of 4 cm. Find the height of each cone. **(Lesson 5-16)**

4. The graph represents the average price of regular gasoline in the United States in dollars as a function of the number of months after January 2014. (Lesson 5-5)

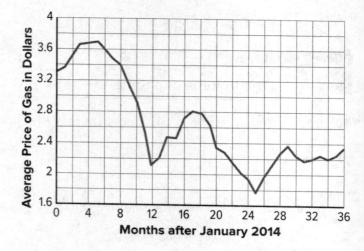

a. How many months after January 2014 was the price of gas the greatest?

b. Did the average price of gas ever get below $2?

c. Describe what happened to the average price of gas in 2014.

5. Match the description of each sphere to its correct volume.

    a. Sphere A: radius of 4 cm   **i.** 288π cm^3

    b. Sphere B: diameter of 6 cm   **ii.** $\dfrac{256}{3}\pi$ cm^3

    c. Sphere C: radius of 8 cm   **iii.** 36π cm^3

    d. Sphere D: radius of 6 cm   **iv.** $\dfrac{2048}{3}\pi$ cm^3

6. While conducting an inventory in their bicycle shop, the owner noticed the number of bicycles is 2 fewer than 10 times the number of tricycles. They also know there are 410 wheels on all the bicycles and tricycles in the store. Write and solve a system of equations to find the number of bicycles in the store. (Lesson 4-15)

Lesson 5-21

Cylinders, Cones, and Spheres

NAME _____ DATE _____ PERIOD _____

Learning Goal Let's find the volume of shapes.

Warm Up
21.1 Sphere Arguments

Four students each calculated the volume of a sphere with a radius of 9 centimeters and they got four different answers.

- Han thinks it is 108 cubic centimeters.

- Jada got 108π cubic centimeters.

- Tyler calculated 972 cubic centimeters.

- Mai says it is 972π cubic centimeters.

Do you agree with any of them? Explain your reasoning.

Activity
21.2 Sphere's Radius

The volume of this sphere with radius r is $V = 288\pi$.

This statement is true: $288\pi = \frac{4}{3}r^3\pi$.

What is the value of r for this sphere? Explain how you know.

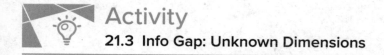

Activity

21.3 Info Gap: Unknown Dimensions

Your teacher will give you either a *problem card* or a *data card*. Do not show or read your card to your partner.

If your teacher gives you the problem card:	If your teacher gives you the data card:
1. Silently read your card and think about what information you need to be able to answer the question. 2. Ask your partner for the specific information that you need. 3. Explain how you are using the information to solve the problem. Continue to ask questions until you have enough information to solve the problem. 4. Share the *problem card* and solve the problem independently. 5. Read the *data card* and discuss your reasoning.	1. Silently read your card. 2. Ask your partner *"What specific information do you need?"* and wait for them to *ask* for information. If your partner asks for information that is not on the card, do not do the calculations for them. Tell them you don't have that information. 3. Before sharing the information, ask *"Why do you need that information?"* Listen to your partner's reasoning and ask clarifying questions. 4. Read the *problem card* and solve the problem independently. 5. Share the *data card* and discuss your reasoning.

Pause here so your teacher can review your work. Ask your teacher for a new set of cards and repeat the activity, trading roles with your partner.

NAME _____ DATE _____ PERIOD _____

Activity
21.4 The Right Fit

A cylinder with diameter 3 centimeters and height 8 centimeters is filled with water. Decide which figures described here, if any, could hold all of the water from the cylinder. Explain your reasoning.

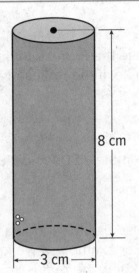

8 cm

3 cm

1. Cone with a height of 8 centimeters and a radius of 3 centimeters.

2. Cylinder with a diameter of 6 centimeters and height of 2 centimeters.

3. Rectangular prism with a length of 3 centimeters, width of 4 centimeters, and height of 8 centimeters.

4. Sphere with a radius of 2 centimeters.

Are you ready for more?

A thirsty crow wants to raise the level of water in a cylindrical container so that it can reach the water with its beak.

- The container has diameter of 2 inches and a height of 9 inches.

- The water level is currently at 6 inches.

- The crow can reach the water if it is 1 inch from the top of the container.

In order to raise the water level, the crow puts spherical pebbles in the container. If the pebbles are approximately $\frac{1}{2}$ inch in diameter, what is the fewest number of pebbles the crow needs to drop into the container in order to reach the water?

Summary
Cylinders, Cones, and Spheres

The formula $V = \frac{4}{3}\pi r^3$ gives the volume of a sphere with radius r.

We can use the formula to find the volume of a sphere with a known radius. For example, if the radius of a sphere is 6 units, then the volume would be. . .

$$\frac{4}{3}\pi(6)^3 = 288\pi$$

or approximately 904 cubic units.

We can also use the formula to find the radius of a sphere if we only know its volume. For example, if we know the volume of a sphere is 36π cubic units but we don't know the radius, then this equation is true:

$$36\pi = \frac{4}{3}\pi r^3$$

That means that $r^3 = 27$, so the radius r has to be 3 units in order for both sides of the equation to have the same value.

Many common objects, from water bottles to buildings to balloons, are similar in shape to rectangular prisms, cylinders, cones, and spheres—or even combinations of these shapes! Using the volume formulas for these shapes allows us to compare the volume of different types of objects, sometimes with surprising results.

For example, a cube-shaped box with side length 3 centimeters holds less than a sphere with radius 2 centimeters because the volume of the cube is 27 cubic centimeters ($3^3 = 27$), and the volume of the sphere is around 33.51 cubic centimeters $\left(\frac{4}{3}\pi \cdot 2^3 \approx 33.51\right)$.

NAME _____ DATE _____ PERIOD _____

Practice
Cylinders, Cones, and Spheres

1. A scoop of ice cream has a 3-inch diameter. How tall should the ice cream cone of the same diameter be in order to contain all of the ice cream inside the cone?

2. Calculate the volume of the following shapes with the given information. For the first three questions, give each answer both in terms of π and by using 3.14 to approximate π. Make sure to include units.

   a. Sphere with a diameter of 6 inches

   b. Cylinder with a height of 6 inches and a diameter of 6 inches

   c. Cone with a height of 6 inches and a radius of 3 inches

   d. How are these three volumes related?

3. A coin-operated bouncy ball dispenser has a large glass sphere that holds many spherical balls. The large glass sphere has a radius of 9 inches. Each bouncy ball has radius of 1 inch and sits inside the dispenser. If there are 243 bouncy balls in the large glass sphere, what proportion of the large glass sphere's volume is taken up by bouncy balls? Explain how you know.

4. A farmer has a water tank for cows in the shape of a cylinder with radius of 7 ft and a height of 3 ft. The tank comes equipped with a sensor to alert the farmer to fill it up when the water falls to 20% capacity. What is the volume of the tank when the sensor turns on? (Lesson 5-13)

Lesson 5-22

Volume as a Function of ...

NAME _____ DATE _____ PERIOD _____

Learning Goal Let's compare water heights in different containers.

Warm Up
22.1 Missing Information?

A cylinder and sphere have the same height.

1. If the sphere has a volume of 36π cubic units, what is the height of the cylinder?

2. What is a possible volume for the cylinder? Be prepared to explain your reasoning.

Activity
22.2 Scaling Volume of a Sphere

1. Fill in the missing volumes in terms of π. Add two more radius and volume pairs of your choosing.

Radius	1	2	3	$\frac{1}{2}$	$\frac{1}{3}$	100			r
Volume	$\frac{4}{3}\pi$								

   a. How does the volume of a sphere with radius 2 cm compare to the volume of a sphere with radius 1 cm?

   b. How does the volume of a sphere with radius $\frac{1}{2}$ cm compare to the volume of a sphere with radius 1 cm?

2. A sphere has a radius of length r.

   a. What happens to the volume of this sphere if its radius is doubled?

   b. What happens to the volume of this sphere if its radius is halved?

3. Sphere Q has a volume of 500 cm³. Sphere S has a radius $\frac{1}{5}$ as large as Sphere Q. What is the volume of Sphere S?

NAME _____ DATE _____ PERIOD _____

Activity

22.3 A Cylinder, a Cone, and a Sphere

Three containers of the same height were filled with water at the same rate. One container is a cylinder, one is a cone, and one is a sphere. As they were filled, the relationship between the volume of water and the height of the water was recorded in different ways, shown here:

Cylinder:

$$h = \frac{V}{4\pi}$$

Cone:

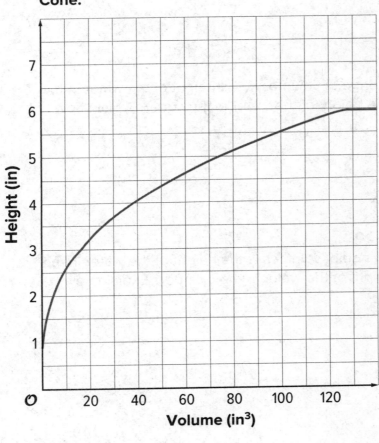

Sphere:

Volume (in³)	Height (in)
0	0
8.38	1
29.32	2
56.55	3
83.76	4
104.72	5
113.04	6
120	6
200	6

1. The maximum volume of water the cylinder can hold is 24π. What is the radius of the cylinder?

2. Graph the relationship between the volume of water poured into the cylinder and the height of water in the cylinder on the same axes as the cone. What does the slope of this line represent?

3. Which container can fit the largest volume of water? The smallest?

4. About how much water does it take for the cylinder and the sphere to have the same height? The cylinder and the cone? Explain how you know.

5. For what approximate range of volumes is the height of the water in the cylinder greater than the height of the water in the cone? Explain how you know.

6. For what approximate range of volumes is the height of the water in the sphere less than the height of the water in the cylinder? Explain how you know.

Learning Targets

Lesson	Learning Target(s)
5-1 Inputs and Outputs	• I can write rules when I know input-output pairs. • I know how an input-output diagram represents a rule.
5-2 Introduction to Functions	• I know that a function is a rule with exactly one output for each allowable input. • I know that if a rule has exactly one output for each allowable input, then the output depends on the input.
5-3 Equations for Functions	• I can find the output of a function when I know the input. • I can name the independent and dependent variables for a given function and represent the function with an equation.
5-4 Tables, Equations, and Graphs of Functions	• I can identify graphs that do, and do not, represent functions. • I can use a graph of a function to find the output for a given input and to find the input(s) for a given output.

(continued on the next page)

(continued from the previous page)

Lesson		Learning Target(s)
5-5	More Graphs of Functions	• I can explain the story told by the graph of a function.
5-6	Even More Graphs of Functions	• I can draw the graph of a function that represents a real-world situation.
5-7	Connecting Representations of Functions	• I can compare inputs and outputs of functions that are represented in different ways.
5-8	Linear Functions	• I can determine whether a function is increasing or decreasing based on whether its rate of change is positive or negative. • I can explain in my own words how the graph of a linear function relates to its rate of change and initial value.

Lesson	Learning Target(s)
5-9 Linear Models	• I can decide when a linear function is a good model for data and when it is not.
	• I can use data points to model a linear function.
5-10 Piecewise Linear Functions	• I can create graphs of non-linear functions with pieces of linear functions.
5-11 Filling Containers	• I can collect data about a function and represent it as a graph.
	• I can describe the graph of a function in words.
5-12 How Much Will Fit?	• I know that volume is the amount of space contained inside a three-dimensional figure.
	• I recognize the 3D shapes cylinder, cone, rectangular prism, and sphere.

(continued on the next page)

(continued from the previous page)

Lesson	Learning Target(s)
5-13 The Volume of a Cylinder	• I can find the volume of a cylinder in mathematical and real-world situations. • I know the formula for volume of a cylinder.
5-14 Finding Cylinder Dimensions	• I can find missing information about a cylinder if I know its volume and some other information.
5-15 The Volume of a Cone	• I can find the volume of a cone in mathematical and real-world situations. • I know the formula for the volume of a cone.
5-16 Finding Cone Dimensions	• I can find missing information about a cone if I know its volume and some other information.

Lesson	Learning Target(s)
5-17 Scaling One Dimension	• I can create a graph of the relationship between volume and height for all cylinders (or cones) with a fixed radius. ◦ I can explain in my own words why changing the height by a scale factor changes the volume by the same scale factor.
5-18 Scaling Two Dimensions	• I can create a graph representing the relationship between volume and radius for all cylinders (or cones) with a fixed height. • I can explain in my own words why changing the radius by a scale factor changes the volume by the scale factor squared.
5-19 Estimating a Hemisphere	• I can estimate the volume of a hemisphere by calculating the volume of the shape I know is larger and the volume of a shape I know is smaller.
5-20 The Volume of a Sphere	• I can find the volume of a sphere when I know the radius.

(continued on the next page)

(continued from the previous page)

Lesson	Learning Target(s)
5-21 Cylinders, Cones, and Spheres	• I can find the radius of a sphere if I know its volume. • I can solve mathematical and real-world problems about the volume of cylinders, cones, and spheres.
5-22 Volume As a Function of...	• I can compare functions about volume represented in different ways.

Notes:

Glossary

A

alternate interior angles Alternate interior angles are created when two parallel lines are crossed by another line called a transversal. Alternate interior angles are inside the parallel lines and on opposite sides of the transversal.

This diagram shows two pairs of alternate interior angles. Angles *a* and *d* are one pair and angles *b* and *c* are another pair.

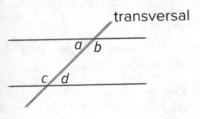

B

base (of an exponent) In expressions like 5^3 and 8^2, the 5 and the 8 are called bases. They tell you what factor to multiply repeatedly. For example, $5^3 = 5 \cdot 5 \cdot 5$, and $8^2 = 8 \cdot 8$.

C

center of a dilation The center of a dilation is a fixed point on a plane. It is the starting point from which we measure distances in a dilation. In this diagram, point *P* is the center of the dilation.

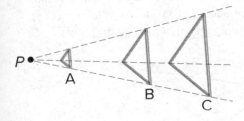

clockwise Clockwise means to turn in the same direction as the hands of a clock. The top turns to the right. This diagram shows Figure A turned clockwise to make Figure B.

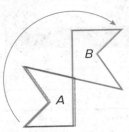

coefficient A coefficient is a number that is multiplied by a variable. For example, in the expression $3x + 5$, the coefficient of *x* is 3. In the expression $y + 5$, the coefficient of *y* is 1, because $y = 1 \cdot y$.

cone A cone is a three-dimensional figure like a pyramid, but the base is a circle.

congruent One figure is congruent to another if it can be moved with translations, rotations, and reflections to fit exactly over the other.

In the figure, Triangle A is congruent to Triangles B, C, and D. A translation takes Triangle A to Triangle B, a rotation takes Triangle B to Triangle C, and a reflection takes Triangle C to Triangle D.

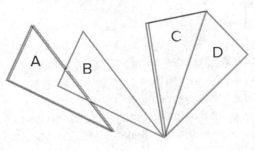

constant of proportionality In a proportional relationship, the values for one quantity are each multiplied by the same number to get the values for the other quantity. This number is called the constant of proportionality. In this example, the constant of proportionality is 3, because $2 \cdot 3 = 6$, $3 \cdot 3 = 9$, and $5 \cdot 3 = 15$. This means that there are 3 apples for every 1 orange in the fruit salad.

Number of Oranges	Number of Apples
2	6
3	9
5	15

constant term In an expression like $5x + 2$, the number 2 is called the constant term because it doesn't change when x changes. In the expression $7x + 9$, 9 is the constant term. In the expression $5x + (-8)$, -8 is the constant term. In the expression $12 - 4x$, 12 is the constant term.

coordinate plane The coordinate plane is a system for telling where points are. For example, point R is located at (3, 2) on the coordinate plane, because it is three units to the right and two units up.

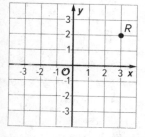

corresponding When part of an original figure matches up with a part of a copy, we call them corresponding parts. These could be points, segments, angles, or distances. For example, point B in the first triangle corresponds to point E in the second triangle. Segment AC corresponds to segment DF.

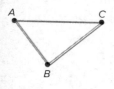
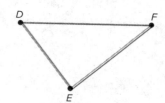

counterclockwise Counterclockwise means to turn opposite of the way the hands of a clock turn. The top turns to the left. This diagram shows Figure A turned counterclockwise to make Figure B.

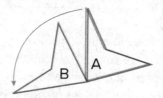

cube root The cube root of a number n is the number whose cube is n. It is also the edge length of a cube with a volume of n. We write the cube root of n as $\sqrt[3]{n}$. For example, the cube root of 64, written as $\sqrt[3]{64}$, is 4 because 4^3 is 64. $\sqrt[3]{64}$ is also the edge length of a cube that has a volume of 64.

cylinder A cylinder is a three-dimensional figure like a prism, but with bases that are circles.

 D

dependent variable A dependent variable represents the output of a function. For example, suppose we need to buy 20 pieces of fruit and decide to buy apples and bananas. If we select the number of apples first, the equation $b = 20 - a$ shows the number of bananas we can buy. The number of bananas is the dependent variable because it depends on the number of apples.

dilation A dilation is a transformation in which each point on a figure moves along a line and changes its distance from a fixed point. The fixed point is the center of the dilation. All of the original distances are multiplied by the same scale factor. For example, triangle DEF is a dilation of triangle ABC. The center of dilation is O and the scale factor is 3. This means that every point of triangle DEF is 3 times as far from O as every corresponding point of triangle ABC.

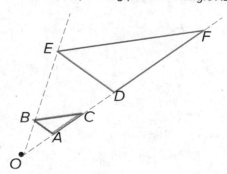

E

exponent In expressions like 5^3 and 8^2, the 3 and the 2 are called exponents. They tell you how many factors to multiply. For example, $5^3 = 5 \cdot 5 \cdot 5$, and $8^2 = 8 \cdot 8$.

F

function A function is a rule that assigns exactly one output to each possible input. The function $y = 6x + 4$ assigns one value of the output, y, to each value of the input, x. For example, when x is 5, then $y = 6(5) + 4$ or 34.

H

hypotenuse The hypotenuse is the side of a right triangle that is opposite the right angle. It is the longest side of a right triangle. Here are some right triangles. Each hypotenuse is labeled.

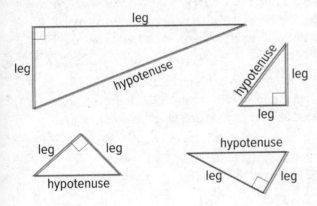

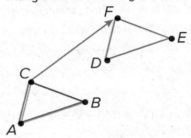

I

image An image is the result of translations, rotations, and reflections on an object. Every part of the original object moves in the same way to match up with a part of the image. In this diagram, triangle *ABC* has been translated up and to the right to make triangle *DEF*. Triangle *DEF* is the image of the original triangle *ABC*.

independent variable An independent variable represents the input of a function. For example, suppose we need to buy 20 pieces of fruit and decide to buy some apples and bananas. If we select the number of apples first, the equation $b = 20 - a$ shows the number of bananas we can buy. The number of apples is the independent variable because we can choose any number for it.

irrational number An irrational number is a number that is not a fraction or the opposite of a fraction. Pi (π) and $\sqrt{2}$ are examples of irrational numbers.

L

legs The legs of a right triangle are the sides that make the right angle. Here are some right triangles. Each leg is labeled.

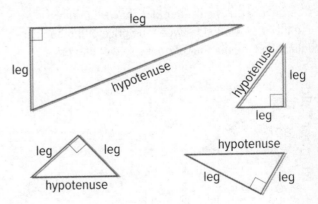

linear relationship A linear relationship between two quantities means they are related like this: When one quantity changes by a certain amount, the other quantity always changes by a set amount. In a linear relationship, one quantity has a constant rate of change with respect to the other.

The relationship is called linear because its graph is a line. The graph shows a relationship between number of days and number of pages read. When the number of days increases by 2, the number of pages read always increases by 60. The rate of change is constant, 30 pages per day, so the relationship is linear.

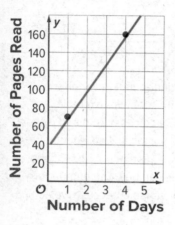

Number of Days

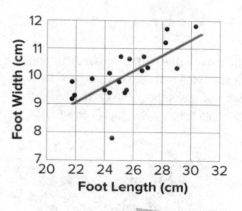

Foot Length (cm)

N

negative association A negative association is a relationship between two quantities where one tends to decrease as the other increases. In a scatter plot, the data points tend to cluster around a line with negative slope.

Different stores across the country sell a book for different prices. The scatter plot shows that there is a negative association between the price of the book in dollars and the number of books sold at that price.

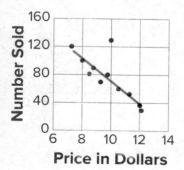

Price in Dollars

O

outlier An outlier is a data value that is far from the other values in the data set. Here is a scatter plot that shows lengths and widths of 20 different left feet. The foot whose length is 24.5 cm and width is 7.8 cm is an outlier.

P

positive association A positive association is a relationship between two quantities where one tends to increase as the other increases. In a scatter plot, the data points tend to cluster around a line with positive slope.

The relationship between height and weight for 25 dogs is shown in the scatter plot. There is a positive association between dog height and dog weight.

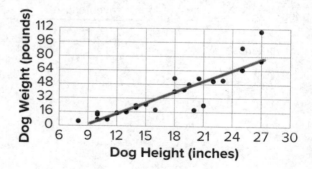

Dog Height (inches)

Pythagorean Theorem The Pythagorean Theorem describes the relationship between the side lengths of right triangles. The diagram shows a right triangle with squares built on each side. If we add the areas of the two small squares, we get the area of the larger square. The square of the hypotenuse is equal to the sum of the squares of the legs. This is written as $a^2 + b^2 = c^2$.

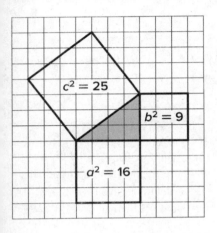

rate of change The rate of change in a linear relationship is the amount y changes when x increases by 1. The rate of change in a linear relationship is also the slope of its graph. In this graph, y increases by 15 dollars when x increases by 1 hour. The rate of change is 15 dollars per hour.

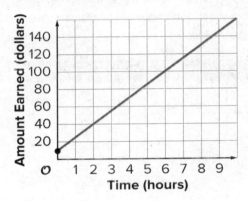

rational number A rational number is a fraction or the opposite of a fraction. Some examples of rational numbers are: $\frac{7}{4}$, 0, $\frac{6}{3}$, 0.2, -$\frac{1}{3}$, -5, $\sqrt{9}$

reciprocal Dividing 1 by a number gives the reciprocal of that number. For example, the reciprocal of 12 is $\frac{1}{12}$, and the reciprocal of $\frac{2}{5}$ is $\frac{5}{2}$.

reflection A reflection across a line moves every point on a figure to a point directly on the opposite side of the line. The new point is the same distance from the line as it was in the original figure. This diagram shows a reflection of A over line ℓ that makes the mirror image B.

R

radius A radius is a line segment that goes from the center to the edge of a circle. A radius can go in any direction. Every radius of the circle is the same length. We also use the word *radius* to mean the length of this segment. For example, r is the radius of this circle with center O.

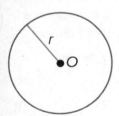

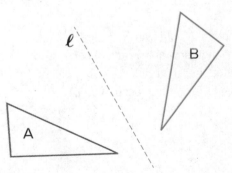

relative frequency The relative frequency of a category tells us the proportion at which the category occurs in the data set. It is expressed as a fraction, a decimal, or a percentage of the total number.

For example, suppose there were 21 dogs in the park, some white, some brown, some black, and some multi-color. The table shows the frequency and the relative frequency of each color.

Color	Frequency	Relative Frequency
White	5	$\frac{5}{21}$
Brown	7	$\frac{7}{21}$
Black	3	$\frac{3}{21}$
Multi-Color	6	$\frac{6}{21}$

repeating decimal A repeating decimal has digits that keep going in the same pattern over and over. The repeating digits are marked with a line above them.

For example, the decimal representation for $\frac{1}{3}$ is $0.\overline{3}$, which means 0.3333333 . . . The decimal representation for $\frac{25}{22}$ is $1.1\overline{36}$ which means 1.136363636 . . .

right angle A right angle is half of a straight angle. It measures 90 degrees.

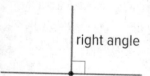

right angle

rigid transformation A rigid transformation is a move that does not change any measurements of a figure. Translations, rotations, and reflections are rigid transformations, as is any sequence of these.

rotation A rotation moves every point on a figure around a center by a given angle in a specific direction.

This diagram shows Triangle A rotated around center O by 55 degrees clockwise to get Triangle B.

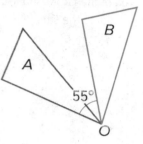

S

scale factor To create a scaled copy, we multiply all the lengths in the original figure by the same number. This number is called the scale factor. In this example, the scale factor is 1.5, because $4 \cdot (1.5) = 6$, $5 \cdot (1.5) = 7.5$, and $6 \cdot (1.5) = 9$.

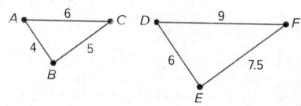

scatter plot A scatter plot is a graph that shows the values of two variables on a coordinate plane. It allows us to investigate connections between the two variables. Each plotted point corresponds to one dog. The coordinates of each point tell us the height and weight of that dog.

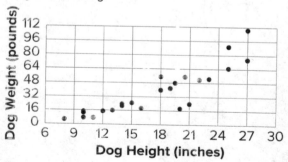

scientific notation Scientific notation is a way to write very large or very small numbers. We write these numbers by multiplying a number between 1 and 10 by a power of 10. For example, the number 425,000,000 in scientific notation is 4.25×10^8. The number 0.0000000000783 in scientific notation is 7.83×10^{-11}.

segmented bar graph A segmented bar graph compares two categories within a data set. The whole bar represents all the data within one category. Then, each bar is separated into parts (segments) that show the percentage of each part in the second category.

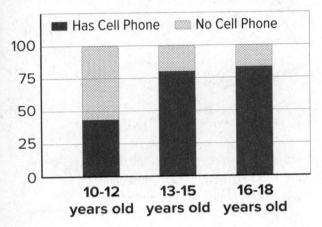

This segmented bar graph shows the percentage of people in different age groups that do and do not have a cell phone. For example, among people ages 10 to 12, about 40% have a cell phone and 60% do not have a cell phone.

sequence of transformations A sequence of transformations is a set of translations, rotations, reflections, and dilations on a figure. The transformations are performed in a given order.

This diagram shows a sequence of transformations to move Figure A to Figure C. First, A is translated to the right to make B. Next, B is reflected across line ℓ to make C.

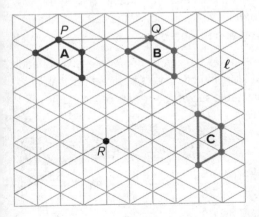

similar Two figures are similar if one can fit exactly over the other after rigid transformations and dilations.

In this figure, triangle *ABC* is similar to triangle *DEF*. If *ABC* is rotated around point *B* and then dilated with center point *O*, then it will fit exactly over *DEF*. This means that they are similar.

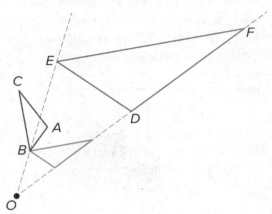

slope The slope of a line is a number we can calculate using any two points on the line. To find the slope, divide the vertical distance between the points by the horizontal distance. The slope of this line is 2 divided by 3 or $\frac{2}{3}$.

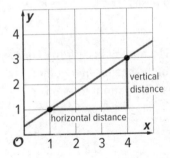

solution to an equation with two variables A solution to an equation with two variables is a pair of values of the variables that make the equation true. For example, one possible solution to the equation $4x + 3y = 24$ is (6, 0). Substituting 6 for x and 0 for y makes this equation true because $4(6) + 3(0) = 24$.

sphere A sphere is a three-dimensional figure in which all cross-sections in every direction are circles.

square root The square root of a positive number n is the positive number whose square is n. It is also the side length of a square whose area is n. We write the square root of n as $\sqrt{n}$. For example, the square root of 16, written as $\sqrt{16}$, is 4 because 4^2 is 16. $\sqrt{16}$ is also the side length of a square that has an area of 16.

straight angle A straight angle is an angle that forms a straight line. It measures 180 degrees.

straight angle

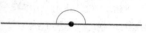

system of equations A system of equations is a set of two or more equations. Each equation contains two or more variables. We want to find values for the variables that make all the equations true.

These equations make up a system of equations:

$$\begin{cases} x + y = \text{-}2 \\ x - y = 12 \end{cases}$$

The solution to this system is $x = 5$ and $y = \text{-}7$ because when these values are substituted for x and y, each equation is true: $5 + (\text{-}7) = \text{-}2$ and $5 - (\text{-}7) = 12$.

term A term is a part of an expression. It can be a single number, a variable, or a number and a variable that are multiplied together. For example, the expression $5x + 18$ has two terms. The first term is $5x$ and the second term is 18.

tessellation A tessellation is a repeating pattern of one or more shapes. The sides of the shapes fit together perfectly and do not overlap. The pattern goes on forever in all directions. This diagram shows part of a tessellation.

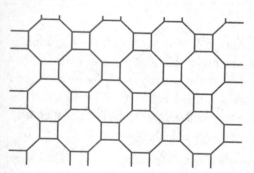

transformation A transformation is a translation, rotation, reflection, or dilation, or a combination of these.

translation A translation moves every point in a figure a given distance in a given direction. This diagram shows a translation of Figure A to Figure B using the direction and distance given by the arrow.

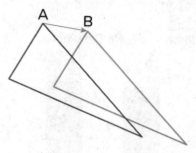

transversal A transversal is a line that crosses parallel lines. This diagram shows a transversal line k intersecting parallel lines m and ℓ.

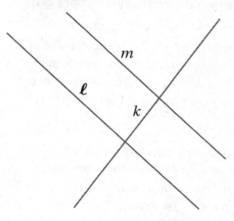

two-way table A two-way table provides a way to compare two categorical variables.

It shows one of the variables across the top and the other down one side. Each entry in the table is the frequency or relative frequency of the category shown by the column and row headings.

A study investigates the connection between meditation and the state of mind of athletes before a track meet. This two-way table shows the results of the study.

	Meditated	Did Not Meditate	Total
Calm	45	8	53
Agitated	23	21	44
Total	68	29	97

V

vertex A vertex is a point where two or more edges meet. When we have more than one vertex, we call them vertices.

The vertices in this polygon are labeled *A*, *B*, *C*, *D*, and *E*.

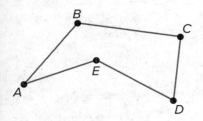

vertical angles Vertical angles are opposite angles that share the same vertex. They are formed by a pair of intersecting lines. Their angle measures are equal.

For example, angles *AEC* and *DEB* are vertical angles. If angle *AEC* measures 120°, then angle *DEB* must also measure 120°. Angles *AED* and *BEC* are another pair of vertical angles.

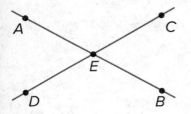

vertical intercept The vertical intercept is the point where the graph of a line crosses the vertical axis.

The vertical intercept of this line is (0, -6) or just -6.

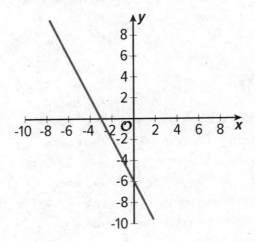

volume Volume is the number of cubic units that fill a three-dimensional region, without any gaps or overlaps.

For example, the volume of this rectangular prism is 60 units³, because it is composed of 3 layers that are each 20 units³.

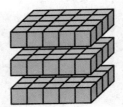

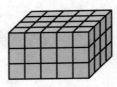

Index